The Public Nature of Science under Assault

Helga Nowotny
Dominique Pestre
Eberhard Schmidt-Aßmann
Helmuth Schulze-Fielitz
Hans-Heinrich Trute

The Public Nature of Science under Assault

Politics, Markets, Science and the Law

Springer

Professor Dr., Ph.D. Helga Nowotny
Wissenschaftszentrum Wien
Strozziggasse 10/16
1080 Wien
Austria
helga.nowotny@wzw.at

Professor Dominique Pestre
Centre Alexandre Koyré
Pavillion Chevreul
Museum National d'Historie Naturelle
57 rue Cuvier
75005 Paris
France
pestre@ehess.fr; pestre@wanadoo.fr

Professor Dr. Dr. h.c.
Eberhard Schmidt-Aßmann
Ruprecht-Karls-Universität Heidelberg
Institut für deutsches und europäisches
Verwaltungsrecht
Friedrich-Ebert-Anlage 6–10
69117 Heidelberg
Germany
schmidt-assmann@uni-hd.de

Professor Dr. Helmuth Schulze-Fielitz
Julius-Maximilians-Universität Würzburg
Juristische Fakultät
Domerschulstraße 16
97070 Würzburg
Germany
schulze-fielitz@t-online.de

Professor Dr. Hans-Heinrich Trute
Universität Hamburg
FB Rechtswissenschaft
Edmund-Siemens-Allee 1
20146 Hamburg
Germany
hans-heinrich.trute@jura.uni-hamburg.de

ISBN-10 3-540-25791-8 Springer Berlin Heidelberg New York
ISBN-13 978-3-540-25791-2 Springer Berlin Heidelberg New York

Cataloging-in-Publication Data
Library of Congress Control Number: 2005929388

Springer is a part of Springer Science+Business Media

springeronline.com

Printed in Germany

Hardcover-Design: Erich Kirchner, Heidelberg

SPIN 11422761 64/3153-5 4 3 2 1 0 – Printed on acid-free paper

Preface

Dieter Grimm

In the period of enthusiasm about technology, the relationship between science and the public was marked by trust. It was taken for granted that scientific progress meant human progress. This is no longer true. Today, it is rather mistrust that dominates. The hope for progress through more knowledge has given way to fear of the risks involved in new technologies in the world of globalized competition. Supervision has replaced the granting of autonomy. The increased importance of knowledge in industrial societies and the great importance placed on science and research have been followed by high demands for accountability and transparency, and have given way to political fights for more money and about the purpose of research. Yet, science insists on autonomy, even if it acknowledges that the public has the right to be informed. *PUSH* (Public Understanding of Sciences and Humanities) is an example of the wide range of efforts to improve communication and to bring science closer to the public. As shown by the sale of popular scientific literature, this effort meets great public demand, though it does not recognizably reduce the conflicts. The question, therefore, is why trust was lost and whether and how it can be re-established.

In the Academic Year 2003/04 an interdisciplinary focus group at the Wissenschaftskolleg was concerned with these problems. The group's intention was to make use of the rich experience in substantive and procedural conflict solution accumulated in the legal science. The turn to law seems to be all the more necessary since trust will no longer be the naive original trust but, given the changes within science mentioned above, a trust of the second order, which results from transparency and the existence of effective safeguards, which rules and law can provide. The law has developed a number of principle-guided procedures for making decisions in situations of uncertainty about their consequences and, in particular, how they can be made when the state is not permitted to decide the question of truth. Also, techniques have been developed that facilitate a revision of previous decisions when better knowledge is at hand.

Some results of the Focus group's considerations are published in this volume. I hope that it will help to establish a trading zone between previously disconnected areas where various concepts must be negotiated, not only within the scientific community, but also with society at large.

Table of Contents

The Changing Nature of Public Science

Helga Nowotny

I. The public nature of science: transformation and new demands

My argument in this introductory chapter is that the public nature of science is changing in a twofold way: one is through the increasing *propertization* of scientific data, what constitutes them, who can have ownership rights over what kind of objects, and what ownership actually means. Propertization is being extended and may now include ownership of novel or existing life forms and living organisms as well as of investigative methods and other scientific procedures. While the overall shift is toward greater regulation of ownership of scientific knowledge, procedures, and results, it follows that the formerly free access and free exchange are shrinking and undergoing various kinds of limitations (Royal Society 2003). Researchers themselves, in order to protect what they regard as their legitimate rights, now often claim individual ownership. In an attempt to defend their autonomy, they see themselves as knowledge owners, rather than knowledge workers (*McSherry* 2001). Unwittingly perhaps, they support the transformation of public science into privately owned knowledge domains.

The second way the public nature of science is challenged and changing carries an apparently different, even opposite message. In the demands for greater public participation in civil society, the public nature of science – in the sense of serving the public good – is no longer taken for granted. Rather, science is challenged publicly as not being public enough. It is obvious that the turbulent relations between science and society do not cover all fields of science and technology, nor is it sheer hostility against science that leads people to question some of its proclaimed benefits. But the fact remains that the most controversial knowledge domains are also the most promising future domains of scientific discovery and technological innovation – biotechnology and biomedicine, nanotechnology, energy production, and environmental issues. This underlines the urgency of finding ways to render new scientific knowledge socially more robust and to integrate future scientific-technological achievements in a culturally and socially compatible way into the lives of citizens. The issues these controversies raise touch some of the core values of modern society and of the modern self – questions of identity, privacy, the alteration of kinship patterns through reproductive medicine, involuntary exposure to risk, and people's relation to the natural environment. Underlying many concerns is the theme of how to preserve a sense of control over one's own life in a bewildering world of scientific-technological complexity, intertwined with the relentlessly ongoing process of globalization.

Public controversies have, for a variety of reasons, led to a pervasive sense of distrust of the independence and impartiality of scientific expertise, side by side with the distrust of political authorities and of industry. The credibility of scientists, even if the public still judges it higher than that of politicians or other interest groups, is in decline. Science is no longer seen as independent and standing above vested interests. In contrast to science's public self-image as neutral, its image has become contaminated by what are perceived as too close ties to state and industrial interests. Moreover, there is an equally pervasive sense that science is not sufficiently listening to citizens' publicly voiced concerns, expectations, and demands. To restore some of the lost credibility and to regain the trust of civil society, so the argument goes, the public should become more involved in the intricate processes of decision-making in techno-scientific issues and developments. However complex these issues may be, if they are likely to have a direct or indirect impact on civil society and therefore will shape the way citizens live and relate to each other, nature, or themselves, the public should be included in the decision-making processes.

In one case to be examined – the question of ownership and of propertization of scientific data – the public nature of science is coming under assault as the result of the real or imagined greater influence of markets and the altered way the demand for increasing investment in research is being met. The extension of the regime of private ownership rights over a growing part of the production of scientific knowledge and its results is coming to be seen as the most efficient form of securing both the needed investment and greater efficiency in producing results. The quest for innovation has extended even to basic research (*Nowotny and Felt* 1997). The push toward more privatization and toward IPR as an efficient form of governance in the domain of knowledge production does not come as an isolated strand. The spreading of affluence in modern societies and the by now ubiquitous presence of modern information and communication technologies have strengthened the tendency to move away from the idea of a centralized state that looks after the needs of its citizens and to move toward satisfying these needs through various forms and degrees of privatization. The rhetoric of the empowerment of the individual, who knows best what is in his or her interest and who masters the art of choice, merely underlines the attractiveness that private ownership – as a promise of individual autonomy – has gained in the context of the interdependencies and complexities of modern societies that are otherwise difficult to penetrate.

In the other case – the demand for democratization of scientific expertise – an increasingly vociferous civil society is questioning the authenticity of the public nature of science. In the demand for greater lay participation, science as an institution comes under pressure to be more accountable to citizens and less closely linked to the interests of politics, the state, and the market. The attractiveness of the shift from the state toward markets and privatization, with their celebration of the individual as consumer and voter, extends to the deliverables that science and technology have been instrumental in bringing to the market or that are offered in semi-privatized public services to the citizens – be they energy, health care, food products, reproductive medicine, or, increasingly, the provision of security. Privatization is therefore not only a powerful underlying theme in the neo-liberal ideology of markets and in political rhetoric, but has also captured the public imagination in the guise of promising greater individual autonomy. The freely choosing consumer is first cousin to the authentic individual (*Skidelsky* 2004).

My argument is that the two apparently opposite tendencies are in fact related. Private ownership and propertization seek to extend a regime that has served industrialized societies well in their pursuit of economic growth. Since science and technology are rightly seen to be the major driving forces of wealth creation and economic growth, some of the latter's governing principles are now expected to work as well in the production of scientific and technological knowledge. The efficiency of markets, competition, and intellectual property rights are to prove themselves by increasing the productivity and output of the production of scientific knowledge, of epistemic things, abstract objects of various kinds, and technological artifacts, all of which constitute the innovative potential of science and technology.

Democratization of scientific expertise is about the extension of principles of governance that also have served Western liberal democracies well in the past. The processes of democratization do not halt before the institution of science. Citizens who have attained a historically unprecedented level of education are no longer over-awed by the achievements of science, presented as miraculous, nor do they accept the word of experts unquestioningly. The process of democratization pushes citizens toward becoming involved in the priority-setting of the research agenda and therefore in the workings of science as an institution that claims to work for the benefit of society. In demanding to re-balance scientific expertise and political representation in the process of actual decision-making concerning scientific-technological innovation and issues, 'society speaks back to science' with a loud and political (sometimes populistic) voice, just as financial investors and markets speak to science when asking for greater efficacy and productivity gains. Calls for accountability and efficacy are the twin approaches that seek to alter the public nature of science as it has developed historically and as we know it today.

The crucial question is, of course, how far the principles of democratic and economic governance under a private property regime can be extended to the actual modes of the workings of science without endangering the autonomy that science will also need in the future – even if ‘autonomy’ and what constitutes ‘science’ may be re-defined as well. The other crucial question is where the visions of a propertized society will lead to and whether such visions and their realization will be politically acceptable. Both the push toward seizing ownership of science and of opening science to demands of democratization must contend with the public nature of science, which they seek to alter in their direction, while science must respond in ways that allow the redefinition of its function while preserving its ways of knowing and working. When science is faced with contradictory pressures from politics, markets, and society, the law can play a role not only in the solution of conflicts, but also in helping to shape a social reality with new institutional arrangements to be put in place. The pressures mirror, and are a part of, far greater economic and political changes that are currently transforming our societies and with which the law should be familiar.

II. The origins of the public nature of science

The institutionalization of modern science in the 17th century depended upon some of its activities being performed in public. What a small group of practitioners, who called themselves natural or mechanical philosophers, did through their investigations of nature, the effects they discovered, and the results they could show was to be displayed in public and ‘witnessed’ in the legal sense of the term. The heterogeneity of the different practices and methods, often subsumed under the generic term of ‘the scientific method’, culminated in scientific experiments. They became the public icon that supported the claims and exemplified the self-confidence of those who were now posing questions to nature and, through the experimental set-up, obliging her to give answers. The crucial act of certifying the results, however, occurs through publication. The process of rendering public through writing, and therefore in traceable, unalterable form, remains crucial to this day. Verification and certification of results and of the methods used is a process that is performed both as an essential service to other knowledgeable members of the scientific community and to the public at large. The peers depend in their own work upon the results of their colleagues. They acknowledge their indebtedness by citing the names of those whose work they have used. All depend upon the quality control that is exercised by the scientific community, through competent mutual criticism and peer review.

Scientists, as a corporately organized collectivity, assume both responsibility for the reliability of their work vis-à-vis each other and public responsibility toward the wider society. In principle at least, scientific findings that have been published should be replicable by others. In practice this is rarely the case, even among specialists, due to the role played by tacit knowledge and other factors. The written form, which lends itself well for certification, and privileges a standardized form of communication, does not sufficiently capture local, practical knowledge, and idiosyncrasies. Knowledge claims are considered temporary, since future findings may at any time overthrow accepted views.[1]

1. Science needs a public – itself

Science therefore is public and depends on having a public – which is first and foremost itself. Publication establishes priority claims and demonstrates the potential usefulness of a researcher's work for others. In recent times, citation counts and impact factors have added to the weight attached to publications as a measure of scientific performance and as a basis for scientific recognition, professional reward, and promotion. Once published, scientific knowledge becomes public and accessible in the public domain, even if it can only be fully understood by other specialists. As long as others can freely use it, science operates as a gift-exchange economy. In return, proper attribution gives credit to those whose work has been used, enhancing thereby their reputation. Although the individual is seen as the originator of new ideas or methods, science operates strongly as a self-organizing collective under a corporatist regime. In this respect, scientists resemble other professionals.

When modern science became institutionalized in the 17th century, the field of natural or mechanical philosophy, as it was called, was populated by amateurs and virtuosi. Practitioners certainly were not the highly professional, formally trained researchers of today. The audience of the Royal Society in London, for instance, before whom demonstrations and experiments were performed, constituted an integral feature of how science was made to be seen 'to work'

---

1 Publication being based upon peer review is so much taken for granted within the scientific community that scientists overlook the fact that this practice is not widely known among the general public. Recently, an advisory panel urged that when scientific results are presented in public, it should be made clear which parts of the claims have been peer reviewed and which ones not, in order to prevent the latter from receiving the same status of credibility in public as peer-reviewed ones. See Nature, 1 July 2004:7.

(*Shapin and Shaffer* 1985). The presence of the public was considered necessary; its function was to 'witness'. Eligibility to become a witness, as in other legal procedures of the time, was limited to those qualifying as 'gentlemen', i.e., men of independent economic means. Others, dependent men and all women, were excluded. Scientific claims had to be made openly, demonstrated, and proven in public. The role of witnessing did not necessarily extend to understand the scientific or mathematical content. It was sufficient to having seen and witnessed what had been told and shown. Thus originated the restricted public of the scientific community – itself – and the wider public before and on behalf of which it acts. Its presence is indispensable to render public what the restricted group of professionals does. In this sense, the ivory tower of an ideal, autonomous science has always been surrounded by 'ivory bridges', linking science to society (*Sonnert and Holton* 2002), while simultaneously retaining its privileged status.

The function of verification, validation, and certification of scientific results is as essential today as it was in the beginning of modern science. The peer group remains the only arbiter believed to be qualified and sufficiently trusted in assessing the production of reliable scientific knowledge. The process of rendering this knowledge public is well known. It begins with the submission of a scientific paper to a professional journal, followed by peer review, usually by anonymous reviewers, and progresses to publication. Today, the style and language of the scientific literature are highly standardized, as is the inclusion of the scientific literature in the citation index. This computerized archive of the scientific literature purports to be international and to represent the most important literature in science and engineering. It is owned by a private company, Thomson ISI, previously known as the Institute for Scientific Information, which indexes more than 8.000 journals in some 30 languages. ISI's monopoly is further strengthened by the measurement of the impact factor, which assigns a weighted number for the frequency of a paper's citation. Although publications are intrinsically linked to the internal reward structure in science, they also function 'externally', e.g. when career recruitment and promotion increasingly are based upon citation measures. The currency of the gift exchange economy, now privately administered, is still the reputation bestowed by the peers in return for the scientific contribution that has been made to the entire scientific community. This currency has not yet disappeared.

The dependence of science on a public, although consisting primarily of other specialists, is therefore essential to the public nature of science. It also sends a strong signal on behalf of a corporate collectivity that it claims collective ownership in the sense of self-regulation. It falls only to the restricted scientific public to certify the claims made in the name of science speaking in a collective voice and to take responsibility toward the wider public. The notion of a restricted public of peers who function as producers and users of knowledge, as collaborators and competitors, as authors and critics of each others' work, is an indispensable condition for the autonomy of science.

Toward the outside, self-governance and non-interference from politics (historically also from religion) is therefore justified by assuming the task of the collective exercise of quality control over its work and results. Utilitarian relevance, which undoubtedly makes support of science attractive to society, is merely one possibility of linking research to societal needs, but does not justify the claim to autonomy, nor does basic research, even though it is highly dependent on being freely and autonomously conducted. The full transparency of obtaining reliable knowledge about nature and for the benefit of society is the reason why misconduct, dishonesty, and fraud are regarded not only as deviant, but also as endangering the claim to the autonomy and self-governance of science. A system that totally depends upon mutual openness, honesty, and trust elicits strong reactions against those who violate these principles, and it must strive to restore credibility as quickly as possible.

2. Secrecy vs. Openness

Another way of potentially undermining both the quality control function of science and its gift exchange economy, occurs through secrecy, i.e. by withholding knowledge from the public domain occurs. Yet, secrecy can officially and legally be permitted in both public and private domains. The predominant reason for secrecy in the public domain is national security. When working for the military or military-related objectives, scientists are usually put under strict rules of confidentiality. Historians of science who have recently reconstructed the conditions for research in the earth sciences in the US during the Cold War period have uncovered an extensive parallel research enterprise, operating side by side with the open science system, although hidden from public view (*Doel* 2003). At present, the US Department of Defense offers contracts to researchers working inside universities that contain not only the usual clauses restricting publication, but also stipulating that research must be shielded from foreign scientists.

The other legally acknowledged exemption from open publication occurs within private industry, especially in the competitive phase of the research and innovation process, when development for market is under way. To protect previously made investments from competitors before products reach the market, publication is restricted or subject to seeking prior permission. This is tolerated in view of the motives of industry and its recognized economic interests (to bring products to market and gain profitable returns on prior investments), just as national security is considered the overriding priority in the case of military-related research.

Imposing restrictions upon the open process of publication therefore occurs both under private and public ownership. Research today is carried out in an increasing variety of institutional settings and is financed from a variety of different sources. In Europe, universities are largely public and publicly financed. In the US, which has many more private universities, universities nevertheless benefit from public tax policies and receive other forms of public support. Research institutions display an extremely heterogeneous organizational structure with some being public, others private, and many mixed. Corporate industrial R&D labs operate under private ownership, often in complex transnational configurations. Small start-up companies engaged in high-tech research and other cooperative ventures exist under a variety of public-private partnerships. The state of public or private ownership is in a condition of extreme flux, with both forms of ownership permitting restrictions of the public nature of science.

Does secrecy endanger quality control? In military-related research this might happen, but for obvious reasons the overall aim is to assure that the quality of military-related science and technology is the same if not superior to research in the public domain. In industry-related research, the interest may be different, especially when negative findings are to be publicly disclosed. In a highly publicized recent case, a British pharmaceutical firm has been sued under New York's consumer-fraud law for 'misrepresenting, concealing and otherwise failing to disclose' the negative results of clinical trials that tested the safety and efficacy of one of its drugs. When contracts are permitted that bind the researchers not to publish without prior permission, vital public interests may be jeopardized. The law is called upon to test the limits of such contractual arrangements (Nature 2004: 589). This case and similar ones have triggered public inquiries, which are to investigate the influence of the pharmaceutical industry on medical research, the promotion of drugs, and regulatory reviews of drug safety and efficacy (*Sismondo* 2004).

Another acknowledged form of restricting the use of the results of research or procedures to obtain them in the public domain is through patents. Typically, patents are granted for limited duration and the protection of invention is conditional upon highly specific criteria, including novelty and inventiveness. Patents remain in the public domain, but access to them so no longer free. The gift exchange economy of science is replaced by a market relations economy.

While the use of the patent or of licensing arrangements is encouraged, it has to be paid for. Patents are usually justified as rewarding previous investment and are regarded as protecting ideas and inventions, precisely by placing them in the public domain. Thus they are seen as reinforcing the public nature of science, while commercializing its use.

3. Speaking in the name of Nature

Since the beginning of its institutionalization in the 17th century, modern science has been public also in another important sense. Its practitioners succeeded in having their claim accepted that they are entitled to speak in the name of a higher authority: Nature. Historically, this claim had to be defended first against established religion and monarchs, and later against the political authority of the nation-state. By agreeing "not to meddle in politicks, rhetoric, divinity…" as the famous phrase of the Charter of the Royal Society stipulated, science had carved out for itself an important autonomous public space. While this autonomy has always been relative, it nevertheless provided the basis and legitimation for demanding the right of non-interference from politics and religion. It lives on to this day under notions like 'Wissenschafts-freiheit', guaranteed for instance in the German constitution, or under the notion of academic freedom, equally protected by law in the anglophone world.

The claim to autonomy is based on a privileged relationship of science to Truth, which is approximated through the procedures and practices scientists use. It is also justified, as shown above, by the service performed for society by assuming collective responsibility for producing reliable scientific knowledge. The ethos of freely sharing results is an integral part thereof. The notion of collective scientific ownership is intimately connected with science's claims to social independence from political and religious powers. Since the ultimate authority to which science appeals, Nature, is above every political authority, and by definition also beyond what can be decided by democratic means and deliberations, science escapes any form of public accountability in its epistemic core and therefore in its 'ways of knowing' (*Pickstone* 2000). At the same time, science is publicly authorized 'to speak truth to power' and therefore to mediate through its special – scientific – expertise between Nature and the exigencies of politics or the demands of society.

The public nature of science thus rests on this dual claim: its special authorization to speak Truth in the name of Nature and to have collective public ownership, including certification and quality control, over the knowledge it produces. Thus at least was the situation as it historically emerged, and these are the claims for the public nature of science that have come under assault today.[2]

III. The contextualization of science: 'society speaks back'

The profound transformation that the science system is undergoing today has led to the development of more open systems of knowledge production, which we labeled Mode 2 in *The New Production of Knowledge* (*Gibbons et al.* 1994). We have also argued that it is becoming more difficult to establish a clear demarcation and differentiation between science and society, since both are subject to the same, or similar, driving forces, among them the inherent growth of new uncertainties and the growing influence of new forms of economic rationality, as well as an increasing capacity for self-organization in both scientific and social arenas. The key to understanding the complex articulations between the social and the scientific as co-evolutionary process is not the impact of any specific parameter, but their suggestive clustering and interdependent influence. There is no metaphysic, no hidden hand, guiding the evolution of science and society in parallel; rather co-evolution is an aspect of coalescence. This does not presume that such processes are in harmony. Quite to the contrary, they can be immensely conflict-ridden (*Nowotny et al.* 2001).

The changing nature of public science is therefore only one facet of many broader changes. But what precisely is changing and where do the pressures for change come from? In *Re-Thinking Science*, my co-authors and I argue for a co-evolutionary process in which science and society become more intertwined. We speak of *contextualization* as embodying a wide range of changes – economic, cultural, political, and social – through which society 'invades' science, but is also 'invaded' by science. In its simplest form, contextualization means that while science has always 'spoken' to society, now 'society speaks back'. Of course, we are aware that these categories are far too crude and encompassing to capture what actually goes on. At the same time, we want to highlight the conditions that enable these mutually invasive tendencies and processes. They are closely related to the shifting boundaries between State and

---

[2] That a government can still seek undue political influence over science is shown by the present US administration. Independence of peer review has repeatedly come under assault, as is the separation of religion and state. Consider the example of the strong tilt of the US administration towards programmes of abstinence rather than more realistic programmes that use contraceptives, to prevent the spread of HIV and other sexually transmitted agents. Second, the policies that now govern stem cell and human embryo research give heavy weight to the moral rights of a tiny cluster of undifferentiated cells at the expense of full fledged adult citizens who could benefit from such research. See *Harold Varmus* (2004) Science, Government, and the Public Interest. APS News.8 June:6.

the Market and between State, Market, Culture, and Science. These boundaries become blurred, occasionally break down, and are reconstituted in a constant flux. The important point is that the process of contextualization influences the conditions under which 'objectivity' arises and is assessed as well as the degree to which the criteria of the reliability of knowledge – which remains indispensable, since science needs to work – are sufficient when dealing, for instance, with issues of risk that have more recently been at the forefront of public controversies.

Contextualization, in the sense of science being open to various influences that shape directions and guide the selection of research priorities, has of course always existed. The dominant context – the normative expectations, economic and political opportunities, and the relation to other forms of cultural, industrial, or military forms of knowledge production – have varied enormously, both in terms of historical time and of geographical and institutional space, as the following chapter by *Dominique Pestre* shows in detail. Science and technology have always engaged with an equally great variety of social actors, forming different kinds of configurations and relationships. Resources and incentives, enticements and affinity to problem choices have played a role, as have local variations of creative scientific and technological hot spots and their embeddedness in networks of exchange and mobility of people and ideas. Although the history has yet to be written of which past scientific and technological developments were financed privately and which publicly, there can be no doubt that the record is extremely varied and mixed here as well.

One major change today is the sheer volume and scale of investment in research and the shift toward an increase from private sources. One can indeed observe a parallel development of the unprecedented expansion of the scope and nature of private financial investment in R&D that went hand in hand with an expansion of what can be privately appropriated, patented, and otherwise propertized. Of course, public investment existing alongside private is nothing new, just as research organizations have displayed great variety, with national, military and industrial goals being pursued quite comfortably alongside intrinsic scientific goals and motivation and often by the same individuals, groups, or institutions. But the shift toward increasing private investment has not only introduced high-tech companies and biotech start-up firms into the stock exchange, but also shareholder values into the management of these companies. They were then passed on to research. Not surprisingly, either, the hot fields of investment are those scientific fields in which advances are most rapid and extremely promising, not only for basic understanding but above all for their potential applications.

Thus, what was once essentially a gift exchange economy among members of the scientific community is being invaded by market relations. The commodification of knowledge, a theme that emerged in the 1970s, has arrived and is here to stay. The public domain is increasingly becoming subject to complex arrangements of legal and financial regulations and restrictions. In the US, already two-thirds of financial investment in R&D come from the private sector, and the European Union is desperately striving to raise its current meager one-third in order to achieve the Lisbon and Barcelona objectives of an overall 3 % of GNP investment in R&D. While an increase of private investment or the spread of various public-private partnerships does not automatically translate into the privatization and propertization of technological and scientific research, the repercussions on the science system are felt throughout. Universities are one crucial site where contextualization or resistance to it is playing out and where the impact is dramatic.

Thus, the context of contextualization has been shaped by a new – at least for Western democracies – climate in which society increasingly speaks back to science. But 'society' is not only civil society. 'Society' also encourages private financial investments when it agrees to decrease public investments and when it supports the creation of an increasingly entrepreneurial environment that invades universities and other research sites. Seen in a longer-term historical perspective, the reality-shaping power of science and technology has become a permanent feature of industrially advanced societies. 'Advancement' and 'progress' can now be achieved only by the continuous and intensive exploitation of the results of scientific and technological research. This process is accelerating – and accelerated by – the dynamics of innovation. Once science could claim to have several, perhaps contradictory functions. Today, its overriding function is to initiate, sustain, and be the main driving force behind innovation. Consequently it is part of the social reality it shapes. The process of contextualization follows the contexts, which multiply.

The extent to which contextualization affects scientific activity, not only in its forms of organization, division of labor and day-to-day practices, in attitudes, values, and ethos, but also deep down in its epistemological core, remains debatable. In *Re-Thinking Science*, we have argued that it does, but many of our critics have – perhaps unsurprisingly – taken issue with our position. What is uncontroversial, however, is the second strong strand of contextualization in which 'society speaks back to science' by demanding to become involved and have a share in political representation when interacting with science.

IV. The demand for the democratization of scientific expertise

When collaborating with industry or striking deals with various influential social groups, when sitting in politically appointed expert commissions, and in many other situations in which close collaboration between state, industry, and science has been a perfectly normal and regular occurrence, the public nature of science takes on a somewhat different meaning from the one that scientists sometimes like to give to it. The heroic gesture of 'speaking Truth to Power' recedes into the background whenever much more mundane, yet urgent administrative and political decisions have to be made. When the issues involved turn around strong scientific-technical dimensions, as they increasingly do in a densely regulated techno-scientific world, scientific expert advice becomes indispensable. Paradoxically, the greater the demand for this kind of expertise, the greater becomes the likelihood of its contestation in the public area.

One reason is that scientific expertise, like other expertise, is inherently transgressive. It must address issues that can never be reduced to the purely technical or solely scientific. Nor does expertise easily fit within the narrow boundaries of an academic discipline. To have any predictive value for action, it must be able to understand the many practical or political linkages that situate and bind the various strands together. It is addressed to decision-makers who seek a sound basis for action and/or legitimation thereof.

Under such circumstances, the claim to speak only in the name of a higher, apolitical authority becomes questionable, especially when it is raised in hotly debated public controversies or when lawsuits against industrial companies put research conducted under its auspices into the limelight of the media. Demands for greater accountability of scientific experts to society will accordingly receive a boost. There are also circumstances in which scientists are seen as behaving just like any other interest group.

1. Expectations and promises

The relationship between 'science' and 'the public' as tension-ridden and characterized by a loss of trust in the disinterestedness of scientific expertise has received a lot of media attention in the past decade. Protests against GMOs, worries about real or imagined risks of new technologies, or displays of outright rejection have become a recurrent feature, highlighting the difficulties of regulating complex techno-scientific issues. It has proved extremely difficult to increase public acceptance of products or deliverables that have come under closer scrutiny as carrying potential risks. Some public controversies exemplify that public expectations were initially triggered by the promises made, either by scientists themselves in the hope that this was a way of attracting research funding or by advertisements or the media. Such promises, when not kept, can backfire. Hype, once uncovered as hollow only serves to further undermine the credibility of those who announce unsubstantiated claims.

Moreover, public expectations have expanded significantly. The technical functioning of a product or the effectiveness of a drug treatment is normally taken for granted. Now the expectation is added that undesirable side effects and unintentional consequences will be minimized, risks made foreseeable and, if deemed unacceptable, completely eliminated. The manifold effects of new scientific-technical products on lifestyle and consumption are touching the formation of social and cultural identities, just as new medical technologies alter people's relationship to their bodies, kinship patterns, and images of life and death. Parents are asking for the right to have a healthy child, soon perhaps with enhanced capacities. The malleability at least implicitly suggested in the promises of further advances also conjures the possibilities of highly individualized satisfaction of human wants. Generally speaking, there seems to be no limit to human desires and how they can be fulfilled with the help of science and technology. The vastly expanded promises that science and technology carry unleash a spiral of further expectations that is difficult to control.

With the waning of the protective and patronizing welfare state, individuals have learned that they must make their own choices, bombarded as they are with the pervasive political rhetoric of choice (*Elam and Bertilsson* 2003). What is often overlooked in the present situation is that the state acted also as a buffer and mediator between science and citizens. It shielded science from too direct demands from the public that were considered inopportune. With the resolute advance of market forces, science also has become more directly exposed to them. Seen from the perspective of a formerly semi-protected science, one kind of exposure is greater dependence on research funding coming from the private sector that translates, for instance, into pressure for more intense university-industry cooperation. Another kind of exposure manifests itself in the demand to engage in much more direct interaction with 'society', although in practice such interaction often takes place through the media. Scientific experts whose work obliges them to deal with a lay public do so by devising highly functional constructs of what my collaborators and I have called 'imagined lay persons' (*Gisler et al.* 2004.). These are abstract entities that are made to fit scientific models and methodologies used by the experts in carrying out their task.

2. Cognitive and normative asymmetries

Confrontations in public controversies involving techno-scientific issues take place in the public arena. They pit scientific experts, who act *in* the name of science but *for* a political or administrative decision-maker, against lay persons who speak in the name of their diffusely defined democratic rights or who claim a diffuse political representation. By definition, experts have knowledge and skills that lay persons lack, so that their relationship is characterized by an epistemic (or cognitive) asymmetry. This inequality is structural. It receives institutional support through the framing of issues, the setting in which the encounter occurs, and the official mandate the experts hold. It is inherent in every interaction between scientific experts and a lay person and forms the basis for the cognitive authority of scientific expertise.

Whenever scientific experts and lay persons are engaged in confrontational interaction, there lurks in the experts a hidden fear: the potential loss of their superior status and their social authority. Since they cannot lose their dominance on the cognitive level, which is structurally assured, the only way lay persons can escape the dominance of the experts is by changing from the epistemic-cognitive regime to a normative-political one. Following a conceptual and empirical distinction introduced by Albert O. Hirschman in the early 1970s when he studied the manifestation of dissatisfaction in organizations, we can distinguish the options of *exit, voice, and loyalty* (*Hirschman* 1970). When lay persons choose the *exit* option, they no longer feel obliged to accept the cognitive authority of scientific experts. The epistemic dominance and cognitive status of experts, which the institutional arrangement guarantees, is questioned and may become irrelevant. In the eyes of lay persons, experts lose their credibility.

Once they no longer feel obliged to be loyal, lay persons have several options. They may retreat into silence and refuse to become engaged. They may opt for protest or they may even seek another, more constructive form of loyalty, by voicing their criticism within the science system and engaging experts in open dialogue. They can also – and this is probably the most frequently chosen option – switch to the political system that offers them *voice,* the recognized possibility to articulate their dissatisfaction. As long as lay persons remain loyal to science and stay within its order, their dissatisfaction does not really matter, since it is not foreseen under the rules governing scientific discourse. Once they have switched to the political system, however, lay persons can confront the experts on normative-political ground.

The institutional arrangements that guarantee cognitive dominance for scientific experts are therefore fragile, since *exit* into the political arena is a legitimate option that becomes attractive under certain conditions. In the political system, lay persons enjoy a different, politically defined status. The balance of power shifts accordingly toward an arrangement in which lay persons' cognitive inferiority to experts disappears. Under the rules of the political system, lay persons can press for laws and regulations that are binding for scientific experts. Parliaments, courts, and administrations, both nationally and on the EU level, have their own experts. They are political, administrative, and legal experts and, in their own domain, *au par* to scientific experts.

3. Institutional arrangements between the scientific and political order

Institutional arrangements regulate the relationship between 'science' and 'the public' on different levels. On the societal macro-level, the relationship between science as a system and the political order is most protective of the public nature of science by constitutionally guaranteeing 'Wissenschaftsfreiheit'. It enshrines the principle that science is speaking truth in the name of Nature and that this is not negotiable and clearly separates science from politics and political interference. On the institutional and administrative level, the options of *exit, voice,* and *loyalty* operate for citizens and curb the cognitive authority of scientists. Experts dominate by definition over the views of lay persons, but this epistemic asymmetry may become suspended when the lay persons exit. Much attention has therefore been given to developing the concept of 'scientific citizenship' (*Felt* 2003). Citizenship is no longer simply taken as given. The newly asked questions are how the varied attributes of citizenship can find expression in societal engagement with science and technology and which purposes are to be served by wider inclusion. The search of new rules of political engagement concerns not only lay people, but governing institutions alike (*Jasanoff* 2004).

In actual controversies, the public nature of science may be contested. The charge is then raised that science is not public enough, because it does not sufficiently take the public interest into account as articulated and represented through the public that is engaged in the controversy. The role of scientific experts who think they know best what citizens need and want is rejected. They are seen as siding with a patronizing State or a profit-hungry industry. Under such circumstances, *voice* becomes a continuous threat to that *loyalty* that citizens have toward science and its experts.

The (relative) retreat of the State and the ongoing project of privatizing economic and social life has the unintended consequence of leaving science (and education) more exposed – and vulnerable – both to market forces and to the demands of citizens as consumers. Accountability can be achieved through markets when political accountability becomes deficient. As science is forced to step outside the shadow of the state, it has to leave part of its privileged status behind. Redefining the public nature of science therefore becomes a double task, since 'the public interest' is to be found in the interpretation given to it by the market or by civil society. Science, when seeking to optimize its self-governance, has to accommodate both.

The greater exposure – and vulnerability – of science to market forces and consumerism, on the one hand, and to the pressures of integrating a number of additional criteria into research and the setting of priorities in the research agenda, on the other hand, manifests itself in the emergence of a permanent conflict-zone of science-society interactions. Some of these conflicts, as *Schulze-Fielitz* points out in his contribution to this volume, are of the moral type that does not permit compromise. While many researchers are willing to subscribe to the expectation that their research will contribute to wealth creation, well-being, and, more generally, to innovation, non-negotiable conflicts are rightly seen as creating barriers for research at the cutting edge. It is therefore important to leave and create space for negotiations. As confrontational as past controversies may have been, for instance in the environmental field, they have led to an intricate and complex, but viable, web of administrative and legal regulations.[3] These prove sufficiently resilient to accommodate research interests, dominant economic interests, the interests of NGOs acting as lobbying groups, the watchfulness of the media, and the political commitment to and/or rhetoric of sustainability. Far from having settled into a comfortable mode of acquiescence, conflicts over environmental issues have tested the limits of what can be negotiated between science, the market, the state, and civil society. These conflicts are not anomalies. Rather, the public nature of science has been re-defined through them.

4. Creating a public space for negotiation

Not all fields of scientific knowledge production or of technology are contested. On the contrary, there is overwhelming consensus – and confidence – that science and technology offer many benefits. Important differences are revealed by surveys of public controversies that have allegedly shaken the

---

[3] See, for instance, the interesting account how US environmental policy early on depended on science, with mixed results. Later, it relied on science to identify problems and solve them, but used other mechanisms, especially the law, to set standards and legal obligations. *Houck, Oliver* (2003) "Tales from a Troubled Marriage: Science and Law in Environmental Policy", Science, December 12, 2003: 1926-1929. Another interesting view on how institutions work is provided by *Thomas Dietz, Elinor Ostrom, and Paul C. Stern* "The Struggle to Govern the Commons", Science, December 12, 2003: 1907-1916.

confidence of citizens. They show that the trust in science and scientists is still way above that in politicians and political institutions, not to mention others (see the chapter by *Schulze-Fielitz,* this volume: 63). They also make clear that the contested issues are not about the laws of nature, nor about Truth in the philosophical sense of the term. Truth has become a far more mundane yet almost unattainable goal: namely, that politicians (or scientific experts) do not lie to the public.

The contextualization of science assures that there is sufficient space for negotiations in the interaction between science and society. This space has to be made institutionally accessible and politically attractive (*Liberatore and Funtowicz* 2003). Once it has become evident that science is not called upon to heroically represent the authority of Nature, the task of creating and shaping spaces for negotiation becomes a more practical one. Institutional arrangements must assure that science can prove itself to be above the private interests of seeking profit and above the interests of political parties and governmental and administrative bodies. It must avoid being seen aligned to economic or political interests, but at the same time should engage with society on a broader base of political representation. The task of redefining the public nature of science can neither escape the political, nor simply defer to it.

The present unease reveals the complex arrangements of a densely regulated world, in which research is entangled with administrative and fiscal regulations, venture capital, and the omnipresent quest for innovation. From the perspective of civil society, science is seen as being deeply implicated in the day-to-day business of providing the market – and society – with ever-new products, new options, and new lifestyles that embody improvements of some sort and the promise of further innovation. Science must come to terms with the fact that, as a consequence of its success in conquering markets, it has lost some of its previous epistemic privileges. Now it must contend with a demanding and fickle public for sustained support.

To provide greater stability, institutional arrangements are called for that lead toward a clearer definition of the rights and obligations of lay persons and that define the space in which negotiations between the epistemic order of science and the normative of politics can be conducted. How this can be achieved will largely be determined in practice. Science must be open to integrating normative commitments into its ways of working and even into its research priorities. The political system has to develop a greater sensitivity to the extent to which self-governance and autonomy of the science system can be combined with its accountability. Some of these the institutional arrangements are already in place, others are under discussion, and some will have to be invented.[4]

The future is inherently fragile and it is not up to science and technology to determine it alone. Modernity began with the intent of shaping what had previously been thought to be beyond human intervention and control. Every major scientific and technological breakthrough with the potential for a large impact on society comes as a package in which social knowledge, knowledge about forms of organization, economic opportunities and constraints, and the potential to empower in shaping the future, are intermixed and intermingle. The process of social and cultural embedding of new scientific and technical advances is as radically uncertain as are the innovations upon which they are based. Civil society has to accept that uncertainties are inherent in the production of scientific and technological knowledge and that not every possible consequence is foreseeable or can be effectively controlled. The extent to which there can be common ground – without a grand vision, but consisting of common projects that convey a sense of social ownership – is to be explored (*Strathern* 2004).

To make such institutional arrangements viable, communication between science and society in the form of dialogue or through other means is certainly desirable, but does not suffice. What must be made visible again is how science and its epistemic order work. In the 17th century, the working of science was demonstrated in front of a public audience. In an age of multimedia and its grandiose means of visualization, this is no longer the issue. What the public should know, and appreciate, is also the social mechanisms through which science works so remarkably well: that criticism and therefore dissent[5] is essential at one stage in the production of scientific knowledge and how consensus is eventually reached, the role played by peer review, that not every statement made in public necessarily represents 'the view of science', how

---

4 One can imagine such an arrangement to include in controversial studies researchers who represent different points of view. Such a proposal was recently put forward by an Editorial in Nature. The controversy is about the respective merits, weighing costs and benefits, of conventional agriculture and methods based on organic farming. The Editorial emphasizes that the available studies are not necessarily flawed. The results often reflect the complexity of the systems being studied, and the number of variables involved. Straight comparisons of organic and conventional practice can often be misleading. Therefore, it would be better to focus studies on particular issues of interest, such as soil conservation, nutrient retention and pest control. These should be done by collaborations between researchers who may have held differing views in the past. See Nature 2004 22 April: 783.

5 See, for instance, the result of a public opinion survey where science is increasingly seen as in internal disagreement about GMOs and cloning. Those who considered the scientific community to be in conflict were also somewhat more likely to be skeptical about biotech applications. *Bucchi and Neresini* 2004: 1749.

misconduct is being dealt with, that fashionable themes also exist within science, how research priorities are being set and how complex the relationships between researchers and the various funding agencies are, that the pressure on younger researchers is mounting to be not only very good in their science, but also to become good entrepreneurs and managers of their labs – and many other themes that are indispensable to an understanding of science as a social, political, economic, and cultural institution. Concentration on the public presentation of the most wonderful, spectacular, and colorful results and images of the stupendous latest achievements of science and technology has been eclipsed the understanding of the processes through which they have been reached, including the social arrangements that make science work. Or, if we want science to be open to the public, we have to beyond showing only the *salon*. To paraphrase Claude Bernard, we have to let the public also into the *kitchen*.

The public visibility of the mode of working of science is a precondition for realistic mutual expectations that will have to be negotiated in the public domain. Other negotiations require an equally careful balancing of, also this time, conflicting interests.

V. Propertization and ownership

"The choices over property forms are choices about the nature of community," writes Peter Drahos, continuing a long tradition of legal and political philosophy that has criticized, justified, proposed, and reflected upon normative theories and institutional arrangements regarding property rights (*Drahos* 1996). Intellectual property rights are an intricate part thereof. Like all property rights, they regulate relations between individuals and not between an individual and things. Unlike other property rights, IPR deals with abstract objects that usually cover technology-based areas like patents and designs or copyright and its cultural subject matter. Abstract objects and the question of who is entitled to access them under what conditions occupy an increasingly important space in the life and functioning of modern society. They are crucial for the ways science and technology work and expand the frontiers of knowledge and artifacts. They create numerous new relationships of interdependence in our societies. Moreover, IPR are closely related to markets and constitutive for the creation and functioning of information markets (*Boyle* 1996). As science and technology advance with unprecedented speed toward ever-new frontiers, there seems to be no end and no limitations to the kind of abstract objects that are created on the way.

At present there is an overt tendency toward the propertization of scientific knowledge and data that formerly were more or less freely available in the public domain or fell under the gift exchange economy of science. Propertization restricts free access and subjects it to various kinds of economic and commercial arrangements. It impacts researchers and the way they do research. Obviously, great variations exist between different scientific fields and some are more directly affected than others. Propertization impacts universities as one of the major sites of knowledge production, all the more since up to now they have been less exposed to commercial considerations than research conducted within industry or in state-owned laboratories. IPR issues are major concerns for universities, both in their dealings with industrial partners and in their internal dealing with members of the university staff. Relationships that were based on trust are increasingly turning into contractual ones. The practical handling of IPR issues shows marked differences between the anglophone world and the European continent, often preventing universities from adopting clear-cut policies. Since taking out patents is not only a complicated, but also a costly process, especially in Europe, costs and benefits need to be carefully weighed against each other.

The tendency to growing propertization is undoubtedly linked to a shift of investment in research from the public to the private sector. While many developments are in flux there is concern, highlighted by the Free Access Movement and the Creative Commons Movement, that cultural and scientific creativity may be stifled if the public realm becomes dominated by property claims and marginalized in relation to the private (*Lessig* 1999, 2004). Moreover, there is an intense ongoing debate about the kind and range of patentable objects, especially when they emerge in the life sciences and in biomedicine (*Davies and Nuffine* 2001). While some of the concerns raised are about possible obstacles to research, others are ethical. The latent fears of the public – to lose control over one's body in the process to be disowned by technoscience – are reinforced by fears that an additional act of disowning may occur when the legal right of ownership over one's body is disappearing. The sense of self and what it means to be human, which is culturally deeply related to images of our bodies and how we relate to them, is undergoing a dramatic transformation as a result of the achievements of genomics, proteomics, and the neurosciences. The ongoing cultural discourse on these changes and their impact reflects the attempt to culturally come to terms with the new scientific realities. The debate on what can be owned adds further disturbing facets to the apparent difficulties in incorporating and appropriating the new knowledge to the changes of identity and the uniqueness of being human.

But choices of property forms are not only choices about the nature of community. Property, as Hegel knew, gives us the means of forming contractual relations with others. To be a person is to be a proprietor, foremost a proprietor of oneself. Property relations are also power relations. In a historical retrospective, which often focuses far too narrowly on developments in a few European countries, private ownership is usually linked to the generalization of a commodity exchange in which everything that can be owned can also be bought and sold. This emphasis on the economic side of property relations paves the way for a generalized mode of exchange through money, capital investment, and returns. The law plays an indispensable role in these developments and is actively involved in the formation of social and political reality. But property relations, far from being absolute in the entitlement they confer to the owner, are also open to negotiations. The role of the law then becomes crucial in redefining which property relations are most appropriate to redefining the public domain and to establish a balance between public and private interests and between markets, politics, and science. Such a balance is all the more difficult to find, since the content and meanings of these categories are changing.

Much of the debate today centers upon the public domain and what ought to remain, be protected, or be placed there. The (re)definition of the public domain is not easy, even if the part we are concerned with, usually referred to as the intellectual or the scientific commons, represents only one, albeit important segment. At first, it seems that the clash is between the interests of private investors, who want their inventions protected through IPR and ask for licensing fees, and public interests as defined by what academics think should remain freely accessible in the public domain. Such a dualistic view is far too simplistic. It does not face up to the fact that IPR has become attractive for many researchers. Although they are not (or not yet) primarily interested in profit, they are attracted by the tactic that the best protection of their research work and intellectual property is through ownership. Even the more general issue of academic freedom is reframed in terms of property rights and no longer seen as an issue of civil rights (*McSherry* 2001). Once the freedom of inquiry is replaced by the freedom of ownership, it is ownership that seems to determine the nature and extent of inquiry.

The (re)definition of the public domain is not easy either, because a broader debate intrudes at every step. The public domain cannot be equated with the public sector, although it includes it. Intermediate bodies that represent various sectoral interests appear and claim to represent the public interest, as do professional groups that partially overlap with researchers and scientific experts. Scientists may act as professionals in the sense that, as in other professions, they are certified, control over entry is in the hands of the profession, and they maintain standards of conduct set by themselves. As professionals, they abide by an ethos of public service. But they also stand up in the public domain to represent their interests, for instance as owners of their scientific work, which they defend against the interests of their employers or colleagues. Universities for their part also have a stake in IPR, partly because patents have become a performance indicator that is increasingly used in the allocation of the public funding they receive and partly because they are regarded (and usually vastly overestimated) as a potential source of future income.

Finally, it is the activity of citizens, and the extent to which they are able to mobilize resources and organize themselves, that forms the basis of claims to define what is in the public interest. Citizens have a general – public – interest in being involved in the definition of what can be patented and how self-ownership of the body or its parts is to be regulated through ownership rights. They have an interest, again of a general and public nature, in participating in the definition and control of information and of what constitute scientific data, since information and a potentially immense amount of data are obtained from them, individually and collectively, on a voluntary or not-so voluntary basis.[6] Genetics has opened up a huge field of scientific inquiry and potential. It is amassing an ever-greater amount of data from citizens. These data are processed and stored, bought and sold, combined with other data and information, and analyzed to obtain highly individualized profiles to assess personal risks, the probability of disease, and other life course events. The debate about propertization in this case and in others joins, echoes, and reinforces civil society 'speaking back' to science as described above.

These overlapping and conflicting interests and claims, rights and contractual relationships, operating under market and non-market conditions, under the shadow of monopolies or in the utopian mood of free access to other users, constitute and populate the public domain. The debate resists being couched solely in legal terms. If left to the law and to the jurisdiction of courts, indispensable as they are, it risks eclipsing the broader, political issues that shape it and that the law also has to consider. It may therefore be useful to revisit the sites

---

[6] Arguing that prospective cohort studies of genes and environment interaction are needed to define the genetic and environmental factors that contribute to health and disease, the author foresees that several hundred thousand people should be enrolled, including a broad age range, the oversampling of minority groups and family-based recruitment that includes multiple generations. Such a study should accumulate, over a considerable time span, not only genetic information, but a variety of environmental influences, leading to the profiling of not only genotypes, but also phenotypes of populations and the single subjects under study. It is easy to see that such a study would raise numerous DNA and other privacy issues. See *Collins* (2004).

where the debate originates and what triggers it, since the nature of the problem may ask for different kinds of solutions.

The first debate is driven by scientific and technological advances. When new knowledge and research technologies come to be at the disposal of researchers, new issues are likely to emerge that did not exist before. Questions may arise – as with the (self-)ownership of genetic sequences – of what can be owned, how far ownership can be extended, with what consequences, and for whom. The creation of new phenomena or objects may become the focus of intense moral conflicts or ethical dilemmas, as in the case of research with embryonic stem cells. Novel and efficient technologies, like microarrays, which permit the analysis and screening of huge amounts of genetic data, necessitate the establishment of new organizations, the formation of public-private partnerships or consortia, and regulations that govern the cooperation between laboratories and the exchange of specimen and of information between research groups.

A second debate, partly triggered by the first and overlapping it, is democracy-driven. Questions about the public interest may pop up in big projects that aim for the collection, analysis, and ownership of genetic and other personal data of an entire population, as was the case with the Icelandic deCode project. Tracking people in comprehensive longitudinal studies and surveys may entail the – highly problematic – proposal that the patients, if they are to receive treatment, may have to give up – voluntarily in advance – some of their rights to privacy, mobility, and even to refuse treatment. Debates may arise over projects of bioprospecting carried out in developing countries that have little or no say and hardly benefit from it, not to mention the complex arrangements in which the WTO is involved through its TRIPS agreement. At stake are not only the technicalities of ownership rights, but political questions of democratic rights, inclusion and exclusion, fairness and justice.

The third kind of debate is essentially institution-driven, whereby the institution is either the university or a research institution whose relation to its staff, to state agencies, or to industry may touch upon questions of the collective ownership of the scientific community. The description and analysis of lawsuits provides ample and instructive material of how far the university has already moved, especially in the United States, in replacing trust by possession (*McSherry* 2001). There is no reason to expect that European institutions, including universities, will not follow suit. The debate frequently aims at reinforcing the need for increased self-governance, with the research or academic community setting its own rules and monitoring their implementation. But as the *Audit Society* has shown, control has shifted from being exercised by an external agency toward a regime combining internal self-discipline with external social control. Audit emerges as "a paradoxical and complex combination of surveillance and trust" (*Power* 1997: 134-5). To the extent that self-governance is an adequate response in the institution-driven debate, nobody should be under the illusion that it will restore the independence of universities or free them from the obligations they enter into, once they accept state or private money.

None of these debates would be as vigorous, vociferous, and volatile in capturing public attention were it not for the media. Following their own interpretation of their mandate to serve the public interest, the media have undoubtedly contributed to holding science accountable and to rendering the changing interface of science and society visible – although they have done so following their own rules of representation.

Five final comments

What follows from this? What is to be done? How much public science does society need, and how public must science be, if both are to prosper in the evolutionary process that binds them together? Five final comments should serve *en lieu* of conclusions.

First, the issues raised in this introductory chapter need an open public debate that places them in a broader, political context. The demand for public participation in a human-built and complex techno-scientific world is understandable, but there are no ready-made solutions to the question of how to balance adequate political representation, scientific know-how and expertise, and the process of decision-making. Controversies between science and the public mirror to a large extent the widespread unease of our political life amid the economic turbulence of a globalizing world in which the instrumental functioning of ownership rights has come to play an important part.

Second, the assault on the public nature of science comes from the twin processes of democratization and the drive for ownership in the form of propertization. In the first case, science is accused of not being public enough and in the latter of becoming too subservient to the private sector. Clearly, a careful rebalancing is needed between the private and the public, the State, its citizens, and the intermediary bodies. By receiving less support from the State, science has become more exposed, both to the market place and to the demands and expectations of citizens.

Third, institutional spaces are needed that function as spaces for negotiating between the conflicting parties and interests. The role of the law is crucial here in devising procedures and setting standards and in following up the working of the institutional arrangements in place.

Fourth, science is deeply implicated in the transformations that touch its public nature and its changing relationship to society. It is important to realize that science cannot 'outsource' its obligations. Rather, ways have to be found to take the societal dimension into science and to integrate it with its ways of working.

Fifth, another way of summarizing the developments sketched so far is to say that we have moved from "Science and Society" to "Science in Society", with its attempts to engage in dialogue and to involve the public through outreach activities. While these should continue, they have not fulfilled the expectation that more information and education will lead to greater acceptance. The time has come to move to the next phase, that of "Society in Science". To safeguard the relative autonomy that science needs, self-governance is a promising strategy. To be successful, it must not stand for closure, but for greater openness. This means science's active involvement in the contradictions and conflicts that are inherent in the life of societies.

Bibliography

Boyle J (1996) Shamans, Software, and Spleens. Law and the construction of the information society. Harvard University Press, Cambridge Mass

Bucchi M, Neresini F (2004) Why Are People Hostile to Biotechnologies? Science 304: 1749

Collins F S (2004) The case for a US prospective cohort study of genes and environment. Nature 27: 475–477

Davies M, Naffine N (2001) Are Persons Property? Ashgate, Darmouth

Doel RE (2004) Constituting the Postwar Earth Sciences: The Military's Influence on the Environmental Sciences in the USA after 1945. Social Studies of Science Social Studies of Science 5: 635–666

Drahos P (1996) A Philosophy of Intellectual Property. Aldershot, Dartmouth

Elam M, Bertilsson M (2003) Consuming, Engaging and Confronting Science. The Emerging Dimensions of Scientific Citizenship. European Journal of Social Theory 2: 233–251

Felt U (2003) Scientific Citizenship. Schlaglichter einer Diskussion. Gegenworte 11: 16–20

Gibbons M, Limoges C, Nowotny H, Schwartzman S, Scott P, Trow M (1994) The New Production of Knowledge: The Dynamics of Science and Research in Contemporary Societies. Sage, London

Gisler P, Guggenhein M, Maranta A, Nowotny H (2004) Imaginierte Laien. Die Macht der Vorstellung in wissenschaftlichen Expertisen. Velbrück Wissenschaft, Weilerswist

Hirschman AO (1970) Exit, Voice and Loyalty. Responses to Decline in Firms, Organizations, and States. Harvard University Press, Cambridge, Mass.

Jasanoff S (2004) Science and citizenship: a new synergy. Special guest editor. Science and Public Policy, 31 (2): pp 2 ff.

Lessig L (1999) Code, and Other Laws of Cyberspace. Basic Books, New York

Lessig L (2004) Free Culture: How Big Media Uses Technology and the Law to Lock Down Creativity. Penguin, New York

Liberatore A, Funtowicz S (2003) Democratising Expertise, Expertising Democracy Science and Public Policy 3: 151–156

McSherry C (2001) Who Owns Academic Work? Harvard University Press, Cambridge

Nowotny H, Felt U (1997). After The Breakthrough. The Emergence of High-Temperature Superconductivity as a Research Field. Cambridge University Press, Cambridge

Nowotny H, Scott P, Gibbons M (2001) Re-Thinking Science. Knowledge and the Public in an Age of Uncertainty. Polity Press, Cambridge (translated into German: Wissenschaft neu denken. Wissenschaft und Öffentlichkeit im Zeitalter der Ungewissheit. Velbrueck Wissenschaft, Weilerswist)

The Technosciences between Markets, Social Worries and the Political: How to Imagine a Better Future?

Dominique Pestre

I have a central aim in this paper – to consider the recent and profound restructuring of science and of society which took shape over the last three decades. This epoch-making transformation has to do with the rules of social life, the dominant norms of justice and of good government, but also, and centrally, with the place, function and usages of the sciences and techno-sciences.[1]

As a way of introduction, I intend to show that the sciences (at least in their 'modern' form, from the 16th Century on) have always been of interest to political and economic powers. As social institutions, they have always been in close relations to various interests and have been produced in a large variety of social spaces – courts, universities, academies, military and engineering institutions, business and popular contexts, etc.

In the second section of this article, I intend to turn my attention to what I take as the first decisive aspect of the new *regime of science production and regulation* that has been established over the last three decades (*Pestre* 2003). I claim we moved from a system of science in society dominated by an equilibrium between science as *public good* and science as *industrial good* to a system in which a financial and market-oriented appropriation of scientific knowledge is now in the ascendant, to science as mainly a *financial good*. This mode of appropriation is both larger in what it includes and rooted in an aggressive extension of property rights. In fact, radically new definitions of patenting, as well as attitudes aiming at invalidating 'commons', were the main tool of this first transformation. In the third and fourth sections, I turn my attention to two other dimensions of that change. First, to the emergence of new kinds of techno-scientific practices and products, notably in the life sciences and infor-

---

[1] This paper would have been impossible without my stay at the Wissenschaftskolleg zu Berlin and the many fascinating discussions I had with Helga Nowotny, Eberhard Schmidt-Assman, Helmuth Schulze-Fielitz and Hans Heinrich Trute, and with Nono (Amnon Raz-Krakotzkin), Gil Anidjar, Peter Hall, Zhiyuan Cui, Susan James, Beate Rössler, Heike Paul and Pascal Grosse.

mation technologies, but also through techniques of modeling, and of mapping at Earth level. I will here insist, however, on the fact that what is usually called our new 'knowledge-based economy' is driven by *a socially-shaped, an economically-embedded arrangement* of science and technology – and not by science-and-technology *per se*. In the following section, I look at the emergence of a new 'civil society' – to employ contemporary parlance – with new relations to industrial techno-science, to moral and social values, and to political regulations. Because the techno-industrial world to which scientific knowledge is now organically linked has the power to dramatically and often irreversibly alter our lives, and because of major transformations in the social body itself, a greater attention and a growing demand for social accountability has surfaced in many segments of the population.

In the last two sections, I consider how such an understanding might help us in devising tools for the future. If we agree on the diagnosis I propose, and if we consider that a lively democracy cannot but rely on different '*cités de justice*' (a notion borrowed from *Boltanski and Thévenot* 1991), certain proposals could be made to help us face the situation which emerged in the 1970s and 80s.

I. Science in Society and Society in Science: A Historical Overview[2]

Let me begin by stating a commonplace: for at least the last five centuries, what we now call scientific knowledge – be it characterized as pure or applied, as natural philosophy or technophysics – has been of crucial interest to the political and economic powers since knowledge led to material and social techniques of control, and to the whole of Europe and the West (as opposed to the Orient) since the new science culturally and morally reshaped Europe, making her (in her own eyes) the only world 'capable' to separate facts from fictions, scientific reason from religion and politics. My second notion, as trite as the first, is that knowledge and science are words that can easily mislead us into inappropriate generalizations if we do not freight them with the precise social and material configurations that obtained to them historically. It is thus essential, for example, to identify where exactly knowledge has been produced. As we know only too well, the fact that Galileo successively worked in a university, then for the Republic of Venice, and finally at the court of the Grand Duke of Tuscany is of direct relevance to the *kind of knowledge* he produced.

2 The ideas behind this section are to be found in the hundreds of science studies' articles and books produced in the last thirty years.

Let me give three examples and try and show how knowledge and social orders were always inextricably entwined; or more precisely, how science-generating activities were not only predicated on social changes but also significantly contributed *to* social change. In the sixteenth and seventeenth centuries, *practical* (or *mixed*) *mathematics* was, for example, a well-recognized field of activity that included astronomy, navigational science, surveying, cartography, geography, fortifications, artillery, the design of mathematical instruments, and the publication of books. Its coalescence at that time was a response to the social, political and economic changes we call the Renaissance, and it directly contributed to its unfolding. Mixed mathematics embraced many kinds of practitioners working for closed or open markets, for state powers or princes, for private entrepreneurs or for companies (like the East India Companies in a later period) and it was advocated as a model for the reform of natural philosophy which took shape in the second half of the seventeenth century and eventually led to what we call modern science. This set of activities did not disappear with the advent of the new natural philosophy, and its heirs could be found in the makers of scientific instruments who supplied the London market in the eighteenth and nineteenth centuries or, for example, in the Army Geographical Services established in late-nineteenth century Europe, generally around geodesy (*Bennett* 1987, out of a very large literature; cf. for details and other references, *Pestre* 2003).

In eighteenth-century Europe we can find another example in the practice of natural history and agronomy. As Emma C. Spary reminds us, our understanding of eighteenth-century natural history as 'pure knowledge' is largely due to our present-day perspective. In fact, it was organically linked with major social projects such as the acclimatization of exotic living organisms, which was seen as having great financial benefits for states, entrepreneurs and peasants; and was linked with the moral reform of society as a whole. Natural history and agriculture were two aspects of the same enterprise aimed at the development of natural resources, the advancement of commerce, and the increase of the nation's wealth. They were advocated and practiced by the same people, both in the *Jardin du Roi* and on private estates (e.g. Buffon's property in Montbard), and the same people served as 'expert consultants', if I could use such an anachronism, to both landowners and the Crown. Natural history and agriculture also provided 'solutions to the problem of the moral and physical degeneration of the nation,' which was of major concern at the time, 'and these sciences came to embody the concerns for social reform of many individuals who were later to be involved in the French Revolution.' Regeneration, liberty and meliorism were at base of many revolutionary acts, 'and they were terms implying a natural and physical process of transformation in living bodies.' Here too we can detect a certain continuity running through nineteenth-century eugenics and clear up to contemporary issues surrounding genetic engineering (*Spary* 2000: 125 and p. 152 for the quotations).

My third example concerns the large systems characteristic of late nineteenth-century and most of twentieth-century techno-science. This covers the whole of chemistry and pharmacy, the bio- and electro-technologies, electronics and the field of material sciences, the control of space and oceans, the production and management of firms, the invention of operational research and system analysis, and so on and so forth. In terms of institutions, this period witnessed the invention of the industrial research laboratory and the establishment of an infinite variety of new relations between universities, state systems, business and the military. It was oriented toward a variety of goals and values (from making money to advancing pure knowledge or fighting communism) and new laws and jurisprudence concerning property rights were devised (basically lasting until the drastic change of the 1980s and 90s). The spectrum of actors became very wide indeed (laboratory-based scientists and engineers, inventors of any kind, amateurs contributing to the advancement of knowledge, scientists devoted to popular education, mathematicians developing tools to rationalize management) and conflicts regularly arose between groups and projects for social, political and epistemic reasons (*Pestre* 2003).

It is not my aim to simply accumulate examples. I only wanted to draw the attention to the fact that science has always mattered tremendously to states and to economic elites; that most science producers have always been attentive to the interests of those elites; and that science has always directly contributed to and been a major resource for social changes and ideologies. I am also saying that each historical moment exhibited a particular combination of these elements, and that a good way to grasp that historical dimension would be to try and describe the various *regimes of knowledge production and regulations in society*, the various regimes of science in society and of society in science that successively or simultaneously took shape. I cannot do that for five centuries, but I can summarize my vision of it for the last 130 years.

What made the century which started around 1870 a profoundly new regime of science and society is that it started a process of *nationalization of science* (*Edgerton* 1997). By nationalization of science I mean that the sciences and their associated technologies have become central to national security, economic development, and the identity of the reshaped nation-states of that period. This has been the case with the ever expanding financing of research and education by states; with the creation of national laboratories beginning with the PTR in Berlin; with large cooperative projects, most of the time centred around prominent companies working in the national (and their own) interest. Although this process of nationalization had its roots at least two centuries back – Colbert's method of managing techno-science and society in France comes to mind – it became emblematic only starting in the 1870s. In fact, because European nations were then at war for global mastery, the state as an institution emerged as the 'natural' centre of society, as the entity around which everything should revolve. The state took upon itself the duty of managing the social peace (becoming a welfare state) and later assumed the task of economic development (becoming a Keynesian state) – largely, if not only, so as to remain an efficient *war*fare *and colonial* state. Via that process, it also became a *techno-scientific* state – this evolution achieving its apex in the professional-administrative state of Cold War America (*Dahan and Pestre* 2004).

II. The System of Science Production and Regulation Today: The New Power of a Financial Markets-Driven Regime

Let me now proceed to the heart of this paper and consider the last three decades. At a macro-economic and political level, I would summarize what everybody knows in the following way: in the past three decades we have moved from systems centrally regulated by democratically elected nation-states to systems without such key central bodies and regulated by authorities that have no direct relation to geopolitical entities, nor to representative democracy. Today, some major regulations are made at the international level through financial markets, through the WTO or the World Bank, through the European Commission and key international companies. In the same move, social beliefs have evolved from strong commitments to social integration via the welfare state to more communitarian and/or individual logics – whatever the varying meaning anybody could give to these expressions (*Falk* 1999 and *Beck* 2002).

This move helped transform the scientific landscape. The universities and their established disciplines have lost their centrality to more inter-disciplinary and inter-*métiers* practices, and to forms of collaboration geared to private demands (think of the Public Private Partnerships now common place in the Health sector in the South). Industrial research is now conceived at international level – the localisation of companies' research centres is now totally open in geographical terms – while universities cannot but remain linked to territories. Interests in science have extended and new institutions have proliferated: venture capital, pension funds, Nasdaq, and start-ups are now part and parcel of the science business (*Nowotny et al.* 2001; *Reich* 2002).

This transformation occurred in response to two impulses. One, in the late 1970s and early 1980s, was the fear of Japanese competition that pervaded all of America. Japan was said to be poised to outstrip the U.S. in new technologies – primarily because the United States had not been protective enough of its fundamental discoveries. The second impulse came from the emergence and very soon the *domination* of finance in the area of industrial activity – including the constant preoccupation with quarterly returns. As a result, drastic cutbacks have affected industrial research since the 1970s (think of the dismantling of the world-famous, century-old, Bell Laboratories, the nucleus of the AT&T empire) and managers placed increasing emphasis on developing more elaborate connections between firms and markets. In short, the dominance of 'market pull' – to employ the local jargon – reasserted itself over 'technology push' and extensive shake-ups took place in most companies' R&D departments (*Buderi* 2000).

More importantly, patent laws and the definition of intellectual property rights were radically reformulated. Let me detail this point. Since the early 1980s in the United States, proprietary rights have been granted on more and more fundamental work, on research which up until then was considered open, public science – and far upstream of any development scheme which alone could be patented. Constraints traditionally attached to patents in terms of precise description of the invention and its usages have been relaxed – which meant the possibility of patenting what was considered science two decades earlier. Patentability has been also opened to laboratory interventions on living entities. That last move started with the patenting of a bacteria in 1980 and of a genetically modified mouse in 1988 and continued up to the patenting of DNA sequences. Today patenting is also common for software, electronic data banks, management tools and so on (*Posner* 2002).[3]

The standard expression to characterize that move in patent regulation is to speak of a new movement of *enclosure*. The analogy is that we face a privatization of the 'commons of the mind' (what public science used to be) which recapitulates, several centuries later, the privatization of the 'common land' in early modern Britain. We know that this movement led to a massive redistribution of wealth, but that it was justified by the new productive possibilities it offered. Some economic historians asserted that this new property regime limited overuse of a land owned by 'nobody' and that it created incentives for large-scale investments. The question is thus to know what to do with this metaphor – whether it is true that public science leaves too much knowledge lying fallow, whether things are always improved by privatization, and who the

3 Criticizing the nearly perpetual patents which can now be taken on business methods, Posner comments: 'Imagine if the first person to think of the auction had been able to patent it.'

losers and the winners might be since a massive transfer of property is never innocent (*Boyle* 2002).

This drastic redefinition of property rights in early 1980s America – your knowledge is now private property – was in part an answer to the fears I just mentioned *vis-à-vis* Japan (to limit free reappropriation by competitors later able to patent their development work). In a recent article, Coriat and Orsi add an interesting element to the picture. They stress that, at precisely the same moment, Nasdaq was also transformed into a stock market for high-tech companies, that pension funds were then allowed to invest in venture capital and Nasdaq companies – in short, that change in patent law was part and parcel of a larger scheme aimed at having fundamental science become first a product on financial markets (*Coriat and Orsi* 2002).

How to assess the consequences of that new set of laws? The debate is still heated but some lessons could be drawn. In that paper, I will limit myself to one particular aspect – the impact *on* the knowledge enterprise. More precisely I will consider if that new patenting regime has favoured innovation – taking for granted that the criteria is that an equilibrium must be maintained between a fair return on investment for the inventor and protection of the common good, which needs as quick a diffusion of the invention as possible. This question is obviously not the only question of importance. As crucial are the social reordering this change in property law implies (what relations are induced by GMOs between peasants and companies, for example?), the consequences for the deployment of culture and free exchange (that is Lessig's point concerning the web), or for an ethics of knowledge policy. The debate between those who favour enclosure and their opponents is thus first and foremost about what is the best 'cité de justice' for our common world (*Lessig* 2004).

About the only question I intend to consider now, the most commonly accepted idea is that if the new equilibrium between private and public science initially gave a push to knowledge production, it might well be that the move went too far. The practice of taking out patents on quite fundamental work may have ended by limiting the number of potential contributors, and by making basic research too expensive (since part of it is patented). This seems to have been recognized by the large companies themselves, particularly in the pharmaceutical industry. Bemoaning the fact that there are too many patents on gene fragments, ten of them chose to recollectivize parts of their data and to make access to them 'free'. In the same spirit, the NIH largely moved against the Bayh Dole Act of 1980 during the second part of the genome project. Their policy was to create large public databases which often led the people who had banked on profit-oriented databases to lower their prices. One could finally understand the feeling of many university science managers who consider the system to be so unwieldy, at least in some sectors, that simpler and more open (non-commercial) relations must be re-established (*Daedalus* 2002).

One could finally anticipate the last two sections by saying that many people in this field refuse the *doxa* that makes what is private better and what is non-private necessarily inefficient. Citing the Linux experience – an open-source software that all could develop for the good of all – they claim that collective work done in a free way is a positive value, that it is more efficient and safer – but also that Linux deserves protection. The question is of importance because this libertarian practice, which started in California in the 1960s, has become an explicit target for large software companies.

III. The System of Science Production and Regulation Today: The Emergence of New Scientific Products and of New Questions

This first story about what might be called 'the commodification of scientific knowledge in the last two decades' is a true and essential one. It is incomplete, however, and one must look at its articulation on other phenomena to better grasp the change. That first transformation in fact took place, at least initially, to integrate the radical changes then affecting technoscience, the tools it mobilized as well as the products it could deliver. The most obvious example is certainly biotechnology's mastery, on the molecular level, of various kinds of living entities. If physicists' mastery of inanimate matter on the atomic and subatomic level started in the 1890s with the discovery and manipulation of the electron, biotechnology on the molecular level is less than thirty years old. More precisely, it started in the second half of the 1970s in the Boston area – and the protection of that emerging field from foreign competitors was certainly decisive in launching the legal change described in the previous section.

The consequences for human beings and societies, however, are to be on another order of magnitude than with physics. No doubt the implementation of micro-physics in nuclear technology triggered extremely important social and political debates, but biotechnology raises far trickier issues. Let me mention three of the numerous issues: (1) The relationship to our own body could well be at stake, a body that biogenetics and surgery might soon transform, reshape, and build at will; this will be a huge step forward that cannot be considered 'scientific business as usual.' (2) The relationship to human reproduction and to filiation is also at stake; reproduction may become largely disembodied, artificial, divorced from women; technology may offer tools for a new kind of eugenics, and the relationship to the definition of life, of the human species and of the social link, may thus be seriously modified – think of cloning.

(3) Finally, our relation to what we used to call Nature will be reversed – with a new relationship to agriculture and food, notably through GMOs, to agricultural societies, particularly in the South, and to the earth as a global living system. In short, these new biotechnologies offer us a way to radically alter our environment and ourselves so much greater in qualitative terms, but also so unknown in its manifold implications, that there would appear to be an element of unbridled *hubris* in it all. And I do not think this impression is merely ignorance or baseless phobia.

Changes are not only happening around laboratory-based sciences such as biotechnology or information techno-science (I have no time to consider here that fantastic and lively field of activity, see *Lessig* 2004). Another major novelty is directly linked to the mathematical and computational means at our disposal to collect, store, and process immense quantities of data, and to the numerous technologies of mapping the Earth – themselves inherited from, and still of the utmost importance to, the most advanced military states. These technical and intellectual means – computers and databases, simulations and models, sensors and techniques on satellites – give us radically new ways of grasping complex and large scale phenomena. The most well-known case is certainly the modeling of climate change which, as everybody knows, includes in its computerized multiple simulations vast mathematical constructions of the atmosphere and of the oceans – but also representations of our own activities as they affect climate change, and an estimate of the effects our economic choices might have on these changes. In a way, these gigantic programs give us a way of 'mastering' the nature/human complex (at least intellectually) but these constructions have no counterweight in society – I mean that there is no equivalent way to apprehend the problem. They tell us what the 'real world out there' likely is, which problems might need an urgent solution, the cost/efficiency ratios we might expect out of our actions – in short, they give a detailed content to feelings we might have but cannot be given an articulated expression (*Demeritt* 2001; *Sarewitz* 2004, on what is at stake in climate change modeling).

Of course, social relations are affected by such developments. Scientific experts are now at the center of many political processes of evaluation and negotiation and ordinary citizens have lost some of their independent judgment. Political authorities have to make decisions based on the trust they put in people in a context of 'uncertainty' – which has led some of them to devise new social and democratic tools like citizen conferences (which also serve, to be sure, in protecting themselves against later reproaches for their decisions). Finally, because stakes are high – if we are to believe most of these simulations, we might very shortly be facing a catastrophic climate change – and because solutions must be implemented *now*, new notions and concepts have emerged in society and in the disciplines of political science and philosophy. Through the concept of sustainable development and the precautionary principle, for example, new ways of thinking the regulation of our activities have recently emerged.

I will finally note – this is the last mutation in the sciences I will mention, even if many others would have deserved a mention – that this move towards predominantly operational values made the notion of science as culture, and of knowledge as personal wisdom, nearly disappear. One key reason is that techno-science is driven by pragmatic attitudes – what counts as good work is 'to get numbers out', as Gerald Holton once said, what is decisive is to train efficient technicians to improve our capability to intervene and technically reshape the world. There is thus a striking contrast with the normative universe that dominated most of the sciences in Europe half a century ago, when 'pure knowledge' was socially valued and when a great scientist was to become a 'savant-philosopher' at the end of his life – think of Niels Bohr. This trend, already quite visible in pre-World War II America, cannot be reduced to the mere logic of industrial techno-science, however. Also important were the maturation of the new tools I have just mentioned: techno-physics and techno-biology, massive flows of data demanding urgent and pragmatic treatment – which led to other ways of working and defining 'science' as an activity; were important the mixing of disciplines and *métiers* all over the industrial technosciences (notably with engineers whose normative principles have always been different from scientists') as well as the fantastic demographic explosion since World War II – which made (techno-) science intrinsically a kind of industry.

Note one central point, however, a point which would deserve far longer a treatment than what I could now do: the ambivalence which *could* be ours in front of such transformations. For two reasons. First because operationality and pragmatism are not uninteresting criteria, far from it, criteria that we could just ignore or dismiss lightly; but also, and more profoundly, because the universal, epistemological and moral, values of science as 'pure' knowledge and culture – notions and values which were invented in several steps from the late XVIIth to the late XIXth Centuries to differentiate 'us-in-the-West' from 'them-in-the-Orient', and 'us-the-scientists' from 'them-the-laypeople' – are not undisputable and without drawbacks. Heavily loaded, 'ideologically speaking', we cannot just accept them as non-problematic.

IV. The System of Science Production and Regulation Today: The Emergence of a New Social and Political Order

Let me now consider a third aspect, essential to understanding today's regime of science in society and of society in science. I mean here the transformations of the social body itself as indicated by the now quite common notion of 'a new civil society', and the parallel transformation of democratic accountability and of the political sphere. This mutation can first be grasped by analyzing the changing composition and geography of our societies (the expansion of new groups with high-level educational capital, the reduction of the blue-collar population, the new forms of inequalities, etc.) and by studying the transformations of social subjectivities and norms (the growth of individualism or of identity politics, the multiplication of NGOs, etc.). New social legitimacies thus appeared on the scene – another reason being that segments of the social body experienced disenchantment with the promise of the national, welfare, planning, developmental and laic state of the post-war and of the decolonization period, in the North as in the South. The global result is a striking upheaval, politically, culturally and socially – as we all know.

To be quick, let me say that the new values first manifested themselves in the late 1960s in the youth and anti-war movements; that they partly consisted in anti-authoritarian attitudes and the rejection of top-down norms, in new sexual and gender relationships, in the claims for an extension and redeployment of the self (*Boltanski and Chiapello* 2000). The consequences are that, today, social perceptions and certainties are not the same anymore. Around science, belief in a self-evident and continuous movement towards progress is eroding – even if that trend is perhaps more discernible in Europe than in America. 'Working for the common good' is also often seen as less attractive than becoming a scientific entrepreneur, and the rationality of science and public action is being contested. Less people believe that technoscience evolves in the right direction; many think that the main regulatory rules concerning industrial technoscience are deficient (as seen in the blood transfusion scandal in France or the mad cow disease epidemics in Europe); expert decisions made behind closed doors are considered undemocratic and participation and control by everyone is claimed as just and fair, normal and necessary.

If I were to put it in another way, I would say that the various '*cités de justice*' (*Boltanski and Thévenot* 1991) which made our common world in the 1940s, 50s and 60s have been profoundly renewed. Let me comment on this statement. What are '*cités de justice*?' They are the common worlds people inhabit, the shared worlds of reference which tell them what is good and what is not. These 'common worlds' are sets of practical and social norms, they are embedded in political and moral philosophies which social actors mobilize to give meaning to their daily activities and decide on what is fair and unfair in concrete situations. There are of course several such common worlds at any given time and people refer to one or another according to the moment or situation.

Freely borrowing from Boltanski and Thévenot, I would characterize the 'common worlds' of the 1950s and 60s, at least in Europe, in the following way: on the one hand, there were three major '*cités de justice*' defining what made a 'good society': (1) the *civic world of the nation-state*, personified by representative bodies like the state apparatus or in the labor contracts signed by worker unions and managers; (2) the *industrial world of enterprise*, interested in efficiency and planning, but also in integrating the workforce and science into the company as a long-term partner; (3) the *hierarchical world of tradition*, the top-down modes of conceiving authority and greatness. Together these three common worlds of social justice defined a dominant way of being in society, stressing respect for labor and merit, for integration and social protection, for progress and the global and rational development of the national group. Then on the other hand, organized around quite different references for justice, was the *market world*, with its appetite for more movement, laissez-faire and opportunistic action, but which still accepted partial state regulation and contractual relations.

Over the last decades, a new 'common world' has emerged, a new set of values for social behaviour and justice – and it has progressively reordered the dominant sense of justice. In this new 'common world', personal engagement, networking and participation, personal responsibility and individual moral behaviour are claimed to be the values that matter, and we are supposed to have moved away from the old dichotomies between 'work and leisure, public and private, profit-sharing and volunteer work, business and ethics'. Derived from that cité are the notion of governance advocated by all European governments and think tanks; business engagements for 'social and environmental responsibility,' so in vogue over the last five years; as well as most actions in the underdeveloped world, when shaped by NGOs funded by the North (*Boltanski and Chiapello* 2000; *Veltz* 2000).

To get a little more precise on that new world of values, I suggest to say some words on the notion of governance, a notion which has largely replaced that of the political in organizing actions in the public sphere. Much has already been written concerning what this word implies, as well as on the chronology of its appearance. The concept was initially elaborated in business management circles at the end of the 1960s when managers were less and less capable of properly running the shop floor – at Fiat, for example, or at General Motors in the United States (*Boltanski and Chiapello* 2000; *Waring* 1991). The concept spread in a first wave at the beginning of the 1980s following the demands for transparency made by managers of pension funds and the renewal of shareholders' involvement in business (*Pestre* 2003). Next, 'good governance' moved into politics via the discourses of international institutions such as the World Bank, and started to appear in the rhetoric of NGOs working in the South to make public life more 'accountable' (*Kalaora* 1999). The final stage of this evolution was the notion of 'global governance,' in part theorized at the London School of Economics, which aimed at establishing common values for the management of a collective, and ultimately reconciled, future. As Moreau Défarges (2001) and others have shown, the vocabulary of governance conveys the idea that the world of the political, as it was invented and has been practiced for decades or centuries, is now obsolete as it is based not only on an overly 'politicized' and conflictual understanding of the social, but also as it relies on the State and formal procedures of representative democracy. As our societies have now been pacified from the outside (we have finished with the Westphalian model and no longer need to fear any wars of aggression – in the North, of course!), and have been pacified from within (we live in wealthy societies and the term exploitation has to be banished) – in sum, because there are no longer any real conflicts of interest or war, and because we no longer have to worry about the redistribution of goods (the free market being the best solution) – the state can fade away and at last allow civil society to auto-organize freely, as it sees fit.

V. What Could be the Rules for a 'Good Society' in Relation to Techno-Science?

I hope you now have an idea of the major changes affecting technoscience in its intrinsic relationship to changing societies today. What I would like to do in the next two sections is to try and imagine the consequences we might be tempted to draw, *for the future*, in more normative terms. The interest of doing such an exercise is twofold. First, I am interested in it as a citizen and as a social critique. Secondly, doing it forces one to precise the description one builds. Thinking in concrete terms and projecting the analysis in the future help explore blind spots.

Before starting, I need some kind of 'reference' for what I would take as a possible 'common good' in social life. Let me start with two quotations.

The first one is from Paul Ricoeur (*Ricoeur* 1991: 166 f.):

> '(...) une démocratie n'est pas un régime politique sans conflits, mais un régime dans lequel les conflits sont ouverts et en outre négociables (...) Sous ce régime, le conflit n'est pas un accident ni un malheur; il est l'expression du caractère non décidable de façon scientifique ou dogmatique du bien public (...) La discussion politique est sans conclusion, bien qu'elle ne soit pas sans décision'.

My second quotation is from Fitoussi (*Fitoussi* 2002: 7 f.):

> 'La démocratie, en empêchant l'exclusion par le marché, accroît la légitimité du système économique et le marché, en limitant l'emprise du politique sur la vie des gens, permet une plus grande adhésion à la démocratie (...) La démocratie, pour que l'économie de marché soit acceptable, doit donc avoir son mot à dire dans les décisions de dévolution des revenus et des richesses'.

These two quotations indicate the direction I intend to follow in normative terms – namely: (1) that it is vital for market democracies – which is a way to define our societies – to keep independent and competing means for the devolution of goods, material goods as well as symbolic ones; (2) that democratic societies rely in an essential way on *not* obeying a single set of norms and on *not* depending on a single *cité de justice*. Equally disastrous would be a world whose decisions would be brought about only by voting procedures, and a world in which market rules would be the only valid ones; (3) that it would be intellectually misleading, pragmatically dangerous, and morally wrong to imagine that it would be feasible and good to take decisions in society based solely on reason and science. Certainly we have to take collective decisions and be informed by all kinds of knowledge, but political discussion and contradictory debates involving ends (What should be the future shape of our society?) and means (What should be the democratic tools?) remain and must remain the key. No science will ever lead to compelling and definitive conclusions, and nothing can replace the expression of conflicting worldviews and common worlds.

Let me adduce a case-study to illustrate this point. The study, conducted by Maurice Cassier and Jean-Paul Gaudillière, concerns the genetics of breast cancer. In their study, the authors identify different (and conflicting) modes of research in this domain along with their concomitant moral and political economies. They show that alternatives do exist and that social choices are indeed at stake at the core of research itself. The first mode they identify is organized around the emblematic and much-admired figure of the scientist-entrepreneur, who depends on the new law regulating the patenting of genes and on the trend toward private insurance companies managing medical care. For the scientist-entrepreneur the best solution (which is simultaneously scientific, medical, and social) would emerge from commercial operations in a free market. The company Myriad Genetics provides here the perfect example: a start-up controlling most of the field through key patents granted on very general grounds, doing research, selling genetic tests and trying to keep competitors at bay.

However, other effective modes of research do exist. One of them relies more upon the association between public research and patient associations (the case of Marie-Claire King at the University of California, who identified the first gene associated with a predisposition to breast cancer in 1990), another on the hospital-based practice of the clinical profession in the context of a public social security system (the case of France, for example). The three modes imply different types of networks and social values, and the Cancer Research Campaign in the United States patented its works to preserve the *public accessibility* of genes against Myriad Genetics. It fought for a reversal of the American patent law on living entities, it contested the constitution of commercial monopolies in the health sector, it petitioned for the establishment of public rules for assessing the clinical utility and the social use of Myriad Genetics tests – in short, it advocated another 'common world', another *cité de justice*. Different arrangements and regimes can indeed coexist, there is no evidence that one mode of production (of knowledge and of society) is intrinsically better and should dominate the others – and the question remains largely political (*Cassier and Gaudillière* 2000).

VI. Some Concrete Proposals

It is time now to take the plunge and try and make concrete proposals. To be as systematic as possible, I should consider the principles and interactions of as many '*cités de justice*' as possible, the way each of them opposes or is ready to compromise with the others. That would be the best way to get a fair idea of what is at stake between them – and thus to create possibilities for reflection and action. This is impossible in so short a text, however, and I will content myself with considering *just two abbreviated examples*. Everybody could then take on the challenge, contest my conclusions, add new points of references, and point to new possible regulatory schemes.[4]

Another word before proceeding. What I suggest we tackle is, I think, larger than what is usually considered in the Science / Technology / Society field under the rubric of dialogic and participative democracy. Dialogic and partici-

[4] This too short a presentation might give the impression I am mainly advocating against States and markets – and so favour NGOs and 'civil societies' actions. That would be too quick an impression.

pative democracy is central and could never be underestimated, but I tend to believe that we have to radically broaden our frames of analysis and consider far more numerous – and *realpolitik*-oriented – social and political regulations. It is in this spirit that I work here – even if this text only considers two particular sets of interactions (*Callon et al.* 2001).

1. The *Cité Marchande* and Its Relation to Other *Cités*

Let me start with the *cité marchande*. If we accept the importance of having competing modes in the allocation of material and symbolic goods, if we accept the importance of defending the autonomy of the various *cités de justice* – and if we consider that there is a growing privatization/financiarization of knowledge today –, working to guarantee the plurality of knowledge-producing institutions, as well as protecting their different moral economies and value systems, seems a good principle *a priori*. It might be translated in one of the following ways:

(1) It is worth defending, at least in the European context, the institutional and financial autonomy of the university structures because the universities are becoming the most dependent link in the (techno-)science system. This defence should not mean a closure of the universities on themselves, however. On the contrary, they should take this new freedom to open themselves to the widest array of social demands. Universities should not align themselves too quickly with market demands, in particular, and academics should follow Daniel Kleinman when he recommends that 'co-operation with citizen groups or NGOs be recognized as part of the service component considered in tenure' when collaboration with industry is considered. If the latter is normal, then why not to have the former become normal too (*Kleinman* 1998)?

This means extending the rights of academics to become full member of the *polis*, to get organical links to civil society as professionals, to multiply their connections with local actors, to imagine solutions which could be simpler and more efficient, technically and socially. More generally, institutions like the French CNRS or the German Max Planck Society should consider their duty to establish structures of concertation enabling them to take on board their programmes, as early as possible, a large set of social questions and worries.

(2) For the institutions whose official function is the protection of the 'public good' (as with the American NIH), this could mean also carefully examining spheres that private interests do not consider. A classic example is chronic diseases in developing countries, another one is the production of knowledge in agro-sciences. As many population geneticists in agronomy will tell you, competition with the large companies of the agro-business is quite unequal.

Monsanto and others have thousands of people working to produce one GMO after another. As soon as they are ready, the pressure is enormous to have these GMOs immediately disseminated throughout the world. On the other hand, there are only a limited number of population geneticists trying to assess the effects these disseminated GMOs might have on eco-systems. If scientific or political institutions consider it useful or prudent to make these assessments before commenting on the possible long-term effects of GMOs, public money should be made available to back this kind of research.

It is thus essential that universities and other academic institutions set up permanent bodies of concertation with as many publics as possible. With economic actors of course – but that is done and is organised through various contractual means. With elected bodies of all kinds – even if the relation of direct dependence must be abolished or, more precisely, must be opened to include other parties. With the whole of the social body, finally – but everything is here to be invented.

(3) A third aspect is what kind of social accountability might be devised for companies other than to its shareholders (which is and will remain, fairly enough, the dominant aspect). The point is central, as techno-scientific products affect us all. Let me take the example of the commercial logic at work in environmental issues. According to Dasgupta, markets are not the most appropriate institutions for protecting the environment. The structural reason he gives is that they 'cannot be relied upon to provide us with prices which would signal true environmental scarcities.' Environmental resources are considered free by most techno-industrial interests (or are consistently undervalued), and there is little incentive to economize their use. If we are to believe Dasgupta, and if we agree that something major is at stake in terms of the 'common good,' it may be essential, for example, to join ecologists and economists in urging public authorities to constitute groups of experts to estimate 'the value of ecosystem services (*Dasgupta* 2000).

In the same way, the question of which institutions should be allowed to define the norms for techno-scientific products is decisive. One example is the *Codex Alimentarius* and the function it has fulfilled since the WTO decided to make it the scientific reference for international trade. Established in 1962 by the FAO and the WHO, the initial role of the *Codex* was to provide minimum safety norms for food; governments and producers, of course, remained free to employ stricter standards. Since its *de facto* redefinition by the WTO, the *Codex Alimantarius* now means that which can be demanded in terms of a product's quality for international trade; stricter rules can thus be declared mere unfair trade practices – and regions or governments which propose tighter regulations on food production (or even explicit labelling on these products, as with GMOs) can be sued. Again, if we believe that science is being used here in an excessive way by the *cité marchande*, then we must find ways of addressing the situation (*Romi* to be published).

Other approaches are indeed conceivable. I will only mention one which started to take shape recently. It is a reaction to the multiplication of 'charters of corporate responsibility' proposed by most large companies nowadays. In these texts, whose social and political functions are quite varied, company managers state what their social and environmental commitments are. Rating agencies (comparable to the ones established on the financial market) have thus started to appear – a famous one in France being established by the former head of one of the major worker trade-unions. Their idea is to assess what companies actually *do* in social and environmental terms, and make their assessments public. They claim that it will help give more weight to these voluntary business commitments, that it will help make these charters *public contracts* (and not only propaganda), that they will make companies more accountable – which is, I think, hardly a naïve idea.

2. The State and the New *Cité Civique*

I now would like to make four groups of remarks regarding states relations to the new civil society initiatives.

(1) Because business and the state can be very closely entwined and both fear interference from 'civil society', I would first stress the importance of protecting the initiatives taken by the new *cité civique*. In terms of science, that might mean protecting the non-governmental research organizations that have been created over the last decades. I have in mind here the case of the CRII-RAD (Commission de Recherche et d'Information Indépendantes sur la Radioactivité), an independent French group of citizens and scientists created in the 1980s which monitors the information provided by France's highly secretive nuclear state-industrial complex. That group proved crucial in its ability to measure radioactivity and then circulate the information. The mad cow epidemics also argue for an active defence of a multiplicity of independent expert sites: all studies have shown, including in Britain, that the techno-administrative milieus in charge of handling the various cases in Britain regularly withheld seminal information – mainly for economical national reasons (*Hermitte and Dormont* 2000).

Another decisive aspect to consider is the conflict of interests in the testing and marketing of new drugs at a time when it has become common practice to have the company itself pay (and use profit-oriented private companies instead of state-governed or academic laboratories) to test the effectiveness and undesirable effects of new drugs. The result of this growing influence exerted by drug companies over the evaluation of their own products is that research is being directed down a narrower path and that the integrity of the clinical research enterprise is being seriously compromised. As two observers recently wrote, 'Drug companies are increasingly funding trials not to discover new agents and new approaches to treatment but to get FDA approval of me-too drugs and to buttress marketing claims.' This desire of companies to control drug assessment also extends to independent research carried out by university departments. It now happens that companies refuse to communicate a new molecule for study before the researcher commits him or herself to not publishing the results without the company's prior approval. This situation has led the high-profile journal The Lancet to take position publicly in its edition of April 14, 2001, contesting the 'right' of any pharmaceutical company 'to block publication of results.' Indeed, it seems that some companies are now ready to sue people who publish without their consent. In Daedalus, Angell and Relman have claimed that after having 'tried to stop publication [of a report] altogether, a company demanded 7 to 10 million dollars in damages [from the group that had been asked to organize the clinical trial of their product and had published the report] on the grounds that publication had hurt the company's financial prospects' (*Angell and Relman* 2002).

(2) A second point regards the organization of formal expertise – since state administrations remain central in the setting up of expert committees. That topic has been thoroughly studied and it seems that real progress has been made in opening up these structures. It is now rather common to have committees that include people representing all concerned parties, to have minorities publicly express their disagreement in expert reports and to have different kinds of review processes (expert committees as well as public hearings, citizen conferences, etc.). What has been won here is the right for non-scientists, non-company engineers and non-state officials to be full members of the process *from the very beginning*, and thus be able to frame the questions right from the start (*Joly* 1999; *Callon et al.* 2001).

Let me be a little more precise and consider first what is always given as the first principle for any proper handling of an expert process: the independence of the expert. In matters of principle (and principles matter tremendously), the idea is crucial and should be emphasized again and again; in the techno-scientific field, however, it is rarely applicable as such, since expertise is only gained by being active in the field – which generally means working for certain interests and not others. We could also worry about the expert's independence vis-à-vis his own discipline's a priori conceptions and prejudices – with its spontaneous scientism for example, or its possible excessive reliance on 'technical' solutions. As an intention or a principle, however, independence is central and the procedures must try to impose at least such an attitude. For example, experts should be asked to publicly declare their links with interested parties; they must be asked to justify their recommendations in writing and publicly; and they should be dismissed if they do not respect these rules (among others *Hermitte* 1998).

A second often-cited principle (more in line with Realpolitik, since it is skeptical about people's good will) is to formally organize the confrontation between experts. Modeled on the system of cross-examination in courts, the method leads to the opening of many closed boxes (or hidden interests) – which is good. But such a systematic confrontation may also prove more dangerous than foreseen if applied on local level. When applied to restricted geographical situations, it could easily disrupt social relationships, exacerbate or unnecessarily polarize positions – and thus obstruct the ultimate goal of any social expertise, which is to facilitate the emergence of a workable compromise. Thus, although this principle is central, especially in formal expert processes on national and international levels, it needs to be implemented, in narrower contexts, in such a way that the expert process does not destroy local relations and could become a learning process (*Hatchuel* 2000).

The third major principle says: do not limit the process to scientists and engineers; have lay people be full members from the beginning; and have them contribute and co-produce expert knowledge directly – for example, in permitting them to have everything revealed at their demand through the creation of panels whose final advice will be highly publicized. The ideas behind this principle are that knowledge benefits by being tackled from various angles; that collective elaboration outside the confines of academic or industrial science is a guarantee of plurality and promises better social outcomes; and that the point of the exercise is not to have a pronouncement of true scientific knowledge and then work to have ordinary people 'understand' it. The implementations of these principles can be seen in consensus conferences, citizen's juries, public hearings, and patient interest groups. Note, however, that each solution leads to different relationships with political or professional authorities or with formal decision-making processes – which remains the most difficult and pressing problem.

I would like to add as a final point the fact that, beyond any formally organized expert evaluation, it might be good to leave some space available for protest. Having accepted that society is not a harmonious whole in which everyone has the same potential to be heard, having accepted that society will never be perfect and that conflict is here to stay, we need to propose a fourth principle: consider protest as both legitimate and normal, consider it as a potential good.

After all, the history of the last few decades has shown how many of the ideas presented in this paper, some of which have already been integrated into our commonsense views, had their origin in protest movements that could not make themselves heard through the conventional channels of social communication. Unless we suddenly become excessively confident in the human race, in society, and in the functioning of institutions, we can take it that this principle will continue to be just as pertinent in the future as well.

(3) That raises, however, one of the more complicated problems – namely the relation to be established between these various kinds of power. Conflicts of democratic legitimacy are unavoidable and there is no ideal prefabricated solution. Elected bodies can claim they are legitimate when defining norms and establishing controlling bodies (e.g. for health or food) but one might claim that democracy is not reducible to these, that other practices compete for legitimacy, and that involving citizens in core decisions is often beneficial. In the matter of techno-scientific products, some governments in Europe have proposed means by which key questions could be publicly assessed and debated. The movement started in Denmark with what has since been dubbed 'citizen conferences.' The question remains, however, as to who has the last word and how to establish explicit reciprocal commitments. If 'citizen conferences' are held, governments should declare beforehand whether and how they intend to act on their recommendations.

The problem becomes even more complex as soon as we address inter-governmental agreements. These are crucial to techno-scientific products and the environment, but very little will come of them if governments do not deal in good faith. Unfortunately, recent events are not reassuring. Two examples are the Montreal and Kyoto Protocols. The first was signed in 1987 and has been in force for fifteen years. At the conference in Nairobi in November 2003, however, the American administration refused continued adherence to the protocol. Since the Montreal Protocol is the only functioning international agreement of its kind, this move was a great disappointment. As for the Kyoto Protocol, we all know that the American administration is not party to it. But if the simulations made by scientists are to be believed, the consequences of not acting now might be terribly damaging. In such context, however, direct means of action are limited.

(4) To conclude – and without becoming too pessimistic – I would like to propose a last comment: decision-making in market democracies depends on a far greater variety of logics *than debate and expertise alone*. Put more bluntly, the sphere of the political is perhaps not *so central* to decision-making because decisions of major importance, in social and environmental terms, are constantly taken 'on markets'. Democratic procedures are often only *reactive*

and occupy a subaltern position with respect to the *de facto* situations created by financial and economic actors. Even though civil society does not lack means to fight back (I could once again evoke the very dynamic 'form of life' of Linux users), acting as though the main regulations were arrived at through participative procedures is certainly too naive.

Echoing Ulrich Beck, we could even say that the main actors of the world economy today constitute a meta-power largely disassociated from the sphere of the political. Their strength resides in their capacity to *do* things, to invest where they deem appropriate, and to pull out of any country that contests their approach. This meta-power is not in need of democratic legitimacy, even if it redefines the political – as moral engagements – and efficiently reconfigures the legal infrastructure – via litigation.

A final question thus arises – and it carries a particular force in the current world where market-based regulations have attained such an amplitude. The question is whether the political, as the set of sites for democratic debate, is still authorized to rule – at least to some extent – on the proliferation of techno-scientific products when their effects are judged by most people to be problematic. The question is whether or not all commerce is justified *a priori*, if the right of commerce could suffer limitations. The tendency today is to say no, that the political cannot and even should not exercise such a power. But this evolution represents a major regression for democracy. After all, for most of the 20th century, the market world 'officially' accepted the possibility of such limitations.

Bibliography

Angell M, Relman AS (2002) Patents, profits and American medicine: conflicts of interests in the testing and marketing of new drugs. Daedalus: 102–111

Beck U (2002) Macht und Gegenmacht im globalen Zeitalter. Suhrkamp, Frankfurt am Main

Bennett JA (1987) The Divided Circle. A History of Instruments for Astronomy, Navigation and Surveying. Phaiden/Christie's, Oxford

Boltanski L, Thévenot L (1991) De la justification, Les économies de la grandeur. Gallimard, Paris

Boltanski L, Chiapello E (2000) Le nouvel esprit du capitalisme. Gallimard, Paris

Boltanski L, Chiapello E (2000) The New Spirit of Capitalism; paper presented to the Conference of Europeanists, March 14–16, Chicago

Boyle J (2002) Fencing off ideas. Daedalus: 13–25

Buderi R (2000) Engines of Tomorrow. Simon and Schuster, New York

Cassier M, Gaudillière JP (2000) Recherche, médecine et marché: la génétique du cancer du sein. Sciences Sociales et Santé 18: 29–50

Callon M, Lascoumes P, Barthe Y (2001) Agir dans un monde incertain, essai sur la démocratie technique. Seuil, Paris

Coriat B, Orsi F (2002) Establishing a New Intellectual Property Rights in the United States, Origins, Content, Problems, Research Policy, vol 31

Dahan A, Pestre D (eds) (2004) Les sciences pour la guerre. EHESS, Paris

Dasgupta P (2000) Science as an Institution: Setting Priorities in a New Socio-economic Context, World Conference on Science, Science for the 21st century, A New Commitment. UNESCO, Paris, pp 264–271

Demeritt D (2001) The Construction of Global Warming and the Politics of Science. Annals of the Association of American Geographers 91(2): 307–337

Edgerton D (1997) Science in the United Kingdom: A case study in the Nationalization of Science. In: Krige, Pestre D (eds) Science in the Twentieth Century. Harwood Academic Publishers, Amsterdam, pp 759–776

Falk R (1999) Predatory Globalization: A Critique. Polity Press, Cambridge

Fitoussi JP (2002) Démocratie et mondialisation. Revue de l'OFCE, hors série: 7–18

Hatchuel A (2000) Agir public et conception collective: l'expertise comme processus démocratique. In: Heurgon E, Landrieu J (eds) Prospective pour une gouvernance. Editions de l'Aube, Paris

Hermitte MA (1998) Pour une agence de l'expertise scientifique. La Recherche 309: 95–97

Hermitte MA, Dormont D (2000) Propositions pour le principe de précaution à la lumière de l'affaire de la vache folle. In: Kourilsky P, Viney G (eds) Le principe de précaution. Odile Jacob, La Documentation Française, Paris, pp 341–386

Joly PB (1999) Besoin d'expertise et quête d'une légitimité nouvelle: quelles procédures pour réguler l'expertise? Revue Française des Affaires Sociales 1: 45–52

Kalaora B (1999) Global experts: la religion des mots, Ethnologie Française XXIX: 513–527

Kleinman D (1998) Beyond the Science Wars: Contemplating the Democratization of Science, Politics and the Life Sciences 16(2): 133–145

Lessig L (2004) Free Culture, How Big Media Uses Technology and the Law to Lock Down Culture and Control Creativity. The Penguin Press, New York

Manent P (2001) Cours familier de philosophie politique. Fayard, Paris

Moreau Defarges P (2001) Gouvernance, Le Débat 115: 165–172

Nowotny H, Scott P, Gibbons M (2001) Re-Thinking Science, Knowledge and the Public in an Age of Uncertainty. Polity Press, Cambridge

Pestre D (2003) Science, argent et politique, un essai d'interprétation. INRA, coll Sciences en Question, Paris

Posner RA (2002) The law and economics of intellectual property. Daedalus: 5–12

Reich MR (ed) (2002) Public-Private Partnerships for Public Health. Harvard Series on Population and International Health

Ricoeur P (1991) Postface au Temps de la Responsabilité, Lectures 1, Autour du politique. Seuil, Essais, Paris

Romi R (to be published) Codex Alimentarius: de l'ambivalence à l'ambiguïté

Sarewitz D (2004) How science makes environmental controversies worse. Environmental Science & Policy 7: 385–403

Spary E (2000) Utopia's Garden, French Natural History from Old Regime to Revolution. The University of Chicago Press, Chicago

Veltz P (2000) Le nouveau monde industriel. Gallimard, Paris

Waring S (1991) Taylorism transformed, Scientific Management Theory since 1945. University of North Carolina Press, Chapel Hill

Comment from a Legal Perspective

Hans-Heinrich Trute

I. Introduction

The "public sphere of science" has many meanings. It can mean the fundamental public character of scientific knowledge, science as the public domain, in which publicizing is the functional precondition for certifying and validating scientific truth claims. But public can also refer to the transparency of the research process and of the science-related decisions of the political or legal system, as well as to the observation of science in the mirror of the media, or to the general public's participation in science, in whatever form and in whatever roles. Dangers can be diagnosed in relation to all these aspects, though for widely disparate reasons and not without ambivalences. If on the one hand dangers to the public through tendencies to secretiveness are impossible to overlook, then on the other hand we find demands for an increased transparency of science-related decisions, for participation in and responsibility on the part of science, as *Helga Nowotny* underscores. These two aspects are connected and can be understood as part of a socialization of science, as a democratization of science (*Nowotny* 2005; *Schulze-Fielitz* 2005), if one does not want to apply this term solely to the political system.

Key words for the changed relationship between science and other sub-systems of society include the economization, politicization, and mediatization of science (*Nowotny* 2005; cf. *Weingart* 2001). The economization of scientific knowledge, in particular, plays a substantial role for the contributions to this publication, if one understands it as a currently especially close coupling between science and business: the orientation of university incentive systems toward the acquisition of third-party funding for specified purposes, cooperation networks between publicly subsidized research and industry, the expansion of the possibilities to patent, secrecy obligations in research contracts, but also dangers to the independence of science through co-optation and ties of interest are some possible aspects of this development. When, in a field like pharmaceutical research, the high-ranking New England Journal of Medicine has great difficulties securing an editorial independent of ties of interest, as its own guidelines demand, and when lack of space prohibits it from publishing the participating authors' conflict of interest statements, then this shows the intensity of the coupling of science and business as if seismographically

(*Angell* 2000 a, 2000 b; *Angell and Relnan* 2002). If we also consider the well-founded recognition that interest ties affect the independence of expertise, then this indicates not only dangers for the public character of science, but also deleterious consequences for other sub-systems that depend on independent scientific expertise, for example, for the political system or legal procedures. The danger of loss of legitimation inherent in the loss of distance, in turn, has effects on other sub-systems and is perceived in the public arena as a loss of independence and trust. As *Dominique Pestre* (*Pestre* 2005) correctly underscores, ties to other societal sub-systems are nothing new in the history of science; but these processes nevertheless reveal an intensity of coupling that, along with other factors, demands scientific attention. The developments merely hinted at here are just one index of the closer couplings with other sub-systems – with politics, religion, and the media, to name only a few.

II. Differentiation or De-Differentiation?

These essays share the hypothesis that the relationship between science and other sub-systems of modern society is undergoing far-reaching changes, but they observe their object from different disciplinary perspectives. Whereas the social-scientific essays thematize the changes, with different nuances in detail, more from the perspective of a de-differentiation of science and society, the normative essays take as their starting point more a differentiation model that, to a certain degree, is given by constitutional law. This does not rule out mutual inspiration in the conclusions to be drawn.

1. Contextualization as De-Differentiation

Helga Nowotny diagnoses the changes against the background of the classical conception, as it were, of the autonomy of science. In the classical model, the public sphere of science is closely connected to the functional mechanisms of the science system as the precondition for the certification and validation of scientific truth claims and for the attribution of reputation (*Weingart* 2001: 68 ff.). At least as science described itself in the 19th century, the public sphere of science is science itself (*Stichweh* 2003). Thus, the science system did not develop the characteristic differentiation between the roles of achievement and audience, which permits the inclusion of virtually all in the audience role, as is characteristic of other societal sub-systems like law, politics, and the economy – or at least not to the same degree (*Luhmann* 1990; *Stichweh* 2003; *Weingart* 2001).

The audience for scientists is initially scientists themselves, so that the achievement and audience roles coincide. Precisely this is recognizable as the specific character of scientific autonomy. Scientific communications initially refer to other scientific communications; they precisely do not answer questions of policy, belief, the law, etc. Thus, the fact that precisely at this time the constitutional guarantees for freedom of science develop can be interpreted as a form of the co-evolution of science and society. Against this background, democratization in the sense described above can be understood as the molding of audience roles, and this not only in the sense of participation, but – where scientific knowledge is commodified – also of customer or client roles. The contextualization hypothesis thus draws attention to the fact that the classical model is currently undergoing changes, especially in regard to science's achievement dimension. Contextualization thus means that the (unforeseeable) implications, as well as the planned or foreseeable applications of scientific research must be taken into account (*Nowotny* 2001: 312) and their consequences taken up, or integrated, in the research process. This is the demand for a reflexive relationship between science and society (*Nowotny et al.* 2001: 69 ff.). Not least because of its success in producing applicable knowledge in ever more societal contexts, science is drawn into ever more societal contexts and thereby simultaneously exposed to competing truth claims, which can trigger intra-scientific needs for reflexivity (*Nowotny et al.* 2001) and which can sometimes be experienced as intra-scientific conflicts. The debate on risk is only one example among many (*Trute* 2005). The contextualization hypothesis thus aims to solve the raised problems through a certain form of integration (for example by including additional stakeholders in the science system), whether in the form of audience or achievement roles. Intensified reflexivity should thus embed science in society (*Nowotny et al.* 2001: 253 f.) by taking society in to science: Society speaks back to science. Connected with this at the same time are changes in the standards and criteria for validating and certifying scientific knowledge, which are expressed in the hope for socially more robust knowledge. This is accompanied by the hypothesis of a de-differentiation of science in relation to other societal sub-systems.

2. Normative Starting Point: A Differentiation Model

If we take this as a greatly simplified version of a model developed in many more facets, then a characteristic difference from jurisprudential observation comes into view. Law necessarily follows a differentiation model in which the autonomy of different social spheres is constitutionally guaranteed. Constitutional guarantees of the freedom of science, as they were instituted initially in Germany and later, especially in the course of the 20th century, more and more frequently in modern constitutions, up to and including the European Charter of Fundamental Rights, European Conventions, and UN Declarations, are – beyond their individualistic dimensions of protecting the individual scientist from state influence – at the same time always also bulwarks protecting a specific social area against de-differentiations. Even where legal doctrines do not state this equally clearly, they can be read this way – also sociologically (*Luhmann* 1965). They provide protection historically, first, against de-differentiations at the hands of the political system and – via the objective-legal dimensions of the fundamental rights (*Schmidt-Aßmann* 2005) – also against societal de-differentiations. In this function, the political and legal system is thus used to maintain a certain degree of autonomy also from societal de-differentiations. This does not prevent either political or societal influence, but create a requirement to justify, which can, but need not be successful.

This is important because the image of science, which *grosso modo* is based on these guarantees, is that it is a *public domain*. Because the public sphere of science is connected with its internal functional mechanisms – with validation, certification, and the attribution of reputation – it is constitutionally guaranteed as public science. In this sense, *Schmidt-Aßmann* can then formulate quite normatively: Science has a distinctly public character. This also means that this fact can be normatively fielded as a shield against de-differentiations at the hands of scientists themselves, as well as against political shapings of the science system's governance structure that seek to use incentive systems to one-sidedly orient it toward the dimension of achievement, thus edging out "freer" research (*Bundesverfassungsgericht* 2004; *Trute* 2005 a). In terms of constitutional law, the state's responsibility for a functioning science system thus also comprises maintaining science as public domain against science's own contracting-out of its public sphere, at least in the area of the publicly financed science system. This constitutional guarantee thus functions also as a symbolic reinforcing of a cultural ideal that definitely has its effects, for example, in the research contracts of major scientific institutions like the Max Planck Society, which are clearly directed against a policy of non-disclosure. The guarantee of freedom thus functions as a buttress and distance-protection and creates requirements to justify (*Schmidt-Aßmann* 2005).

This should not be misunderstood as a petrifaction of a certain image of science that could be brought into position against changes in scientific practice. Thus, there is no objection to including stakeholders in the research process; rather, this is a component of freedom itself, which is also always freedom to make changes, as long as they are not coerced. Nor is the state prevented from applying other constitutionally protected legal positions and from implementing them in demands on the science system, i.e., ensuring and shaping rights of access to scientific knowledge; and this also when it is generated in the publicly financed part of the science system. Even less is the state prevented from applying other societal claims in decisions on structures of the research system or on funding priorities or in giving processes a shape that permits these claims to be applied. All of this at least as long as it does not endanger the functioning of the science system in the sense that free research would no longer be possible.

3. Normative Conclusions for Science in Contexts

But despite all normative flexibility, the basic tension between social-scientific observation of processes of de-differentiation and legal concept remains, of course. In a differentiated society, plural description perspectives are as unavoidable as are differing scientific perspectives. Even if legal guarantees always refer to and thus also change along with a social reality whose normative order they aim at, the *artificial reason of law* prevents any unrestrained snuggling up to the science system's theories of reflection. Normativity cannot be had any other way, since it is always also contrafactual. It guarantees the security of expectation even when the world changes. Formulated sociologically, therein lies its societal function. Of course this does not prevent a co-evolution of law and social reality, as we saw above in the crystallization of the guarantee of scientific freedom. Thus, this tension points to the necessity to scientifically examine the adequacy of the image of science used by the law.

But in respect to this, there are characteristic difficulties, including those of interdisciplinary dialog. In its decisions, the *legal system* necessarily narrows the plurality of scientific descriptions – legal and social-scientific constructions of the diagnosed changes – deciding for one viewpoint, though it could of course choose a completely different one. The (heterogeneous) scientific descriptions of empirical reality and of the law are thus offers to the legal system to see something in a certain way. The way the legal system sees the world can then be deduced from its decisions. These, of course, cannot simply accept specific scientific descriptions (which of the competing descriptions should they accept?). Rather, the descriptions are selected and modified in the framework of the legal system's own context of argumentation and reference (what precedent decisions, what developments in the law, what standards, etc.) They are thus necessarily normative constructs, including of reality, whose order they aim at. This is what is meant by *artificial reason of law*, which embodies the achievement of modernity in that it separates legal decisions from scientific, political, and moral aspects without questioning the legitimacy of the latter and without denying the contingence of legal decisions (everything can also be decided differently). But also without neglecting mutual dependencies and influences. This, in turn, can be observed jurisprudentially and social-scientifically. In interdisciplinary dialog, we will thus have to distinguish more precisely whether we are speaking of (legal-) scientific descriptions (of the law or of reality) or of legal decisions. But it is definitely a matter of constructs that – depending on the standpoint – could always also be different. So legal scientists are thus the last ones who would be surprised by the scientific diagnosis that the world (of the law or of reality) can also be seen quite differently. It would be much easier to surprise them with the hypothesis – because from their viewpoint it is contrafactually formulated – that de-differentiation should include to relinquish the respective rationalities of law, politics, science, etc., however these should be described. But if de-differentiation means the mutual coupling of various sub-systems and the changes in their forms, about which there can be no serious doubt, then these can be reconstructed in various theoretical models – social-scientific as well as jurisprudential models, incidentally. That a normative starting point thereby chooses a differentiation model is hardly surprising, after all this.

The real challenge on the legal-scientific level, then, consists in developing, through interdisciplinary dialog, normative requirements for the changed conditions of the production of scientific knowledge. In this sense, the legal-scientific essays display approaches oriented toward a normative constitution of these changed forms of knowledge production. Changed legal-scientific descriptions of science, the emphasis on the plurality of scientific expertise, demands on the independence of science in political processes and normative procedures, demands for the transparency of procedures, and the publication of results – in relation to contextualization, all of these can be seen as part of the attempt to formulate a new normative framework for science in contexts – as offers to the legal system and as offers in scientific discourse.

4. Contextualization and Juridification

But at the same time, this also makes it clear that, jurisprudentially, a number of developments reflected in the contextualization hypothesis are not problems of internal change in science's relationship to society, but a coupling of divergent sub-systems, typically science, business, politics, law, and the media. The consequence, of course, is also a juridification of the context. The more the context is integrated in science, the closer the ties between science and other sub-systems also become. The consequence of this close coupling is, then, also an increasing degree of juridification (*Schulze-Fielitz* 2005) of science. This is thus an expression (and gauge) of contextualization. This is not only a shift in the difficulty of inclusion in the sense just described, which becomes an inclusion in the legally codified procedure even in regard to participation in political decisions. The object and processes also change. Including stakeholders in the research process may, as Nowotny describes, alter the result as *socially more robust knowledge* is produced; it also makes it clear that the regulation context is differentiated not only in terms of the matter at hand (cf. the typology of *Schulze-Fielitz* 2005). It is also highly differentiated nationally or territorially. Diagnoses of development may prove to be more ambivalent than suspected, as shown in Schmidt-Aßmann's essay on the various approaches to patent law in Europe, Germany, and the USA. Differing conceptions of science, property, precautions, etc. influence the approach to regulation. Which approach finally prevails is doubtless a question of political and economic power, but precisely for that reason also an empirically open question that attaches reservations to possible generalizing diagnoses of developments.

5. Differentiation of the Inclusion Problematic

If, as the essays by *Schulze-Fielitz* and *Trute* show, some of the discussed problems in the relationship between science and the public sphere are better grasped as problems of including the public in the political or legal system, in which science is either the object of decisions or provides knowledge for political or legal decisions, then these decisions also require other normative rules. Thus, where inclusion, transparency, and the public sphere respectively begin or should begin must be made more specific from the perspective of the legal system. Many of the demands for a greater role for the public sphere address the political system in its science-related or science-based decisions, for example, in setting priorities or in deciding what will be funded or risked, and present themselves as inclusion in administrative processes or the political process. A number of the initiatives to evaluate technology, like consensus conferences, planning cells, and focus groups, do not aim at inclusion in the science system, but definitely belong to the run-up to political decisions. They may – apparently rarely enough (*Abels and Bora* 2004) – change the political process toward more responsiveness to societal plurality, absorb protest, and in this sense be understood more as a democratization of politics. But they affect science only through the result of political decisions, which may be transformed in law. The same is true for decisions about risk, if these are not at the same time the risks of research, for example in part the decisions about releasing genetically modified organisms. Here, too, it is becoming clear that the law is used to search for greater differentiation that would enable a clearer normative separation between scientific, legal, and political questions (*Trute* 2005). What is coming to view in these fields is a complex, normatively structured arrangement designed to ensure that the recognizable tendencies toward the politicization of science are countered by the creation of separate arenas. Here, the law has a differentiating effect and, in a certain way, functions to protect distance. Thus, the legal findings of the EuGH in the field of risk law make it clear that demands on science are formulated in regulatory contexts that work against the tendencies toward a loss of influence of the public sphere

(*Trute* 2005) and that connections to the public science system are formulated (transparency, excellence). This includes roles for the public in the generation of knowledge or enables such roles to be planned. Thus, here normative arenas are established in which *socially more robust knowledge* is, or at least can be, generated.

This at the same time makes it clear that the legal system very specifically shapes inclusion and (drawing on a wealth of experience) does not necessarily provide for an integration through participation (*Trute* 2005). Here there are clear parallels to the perspective of Dominique Pestre, who presupposes that many of the relevant questions of techno-science are above all political questions that must be processed also in the political-administrative system and that thus have to do with the configuration of the political realm. In itself, the law is flexible enough to work through the conflicts that arise here, but it cannot compensate for the deficits of the political process without being overburdened with political claims in administrative procedures. But the law can formulate normative demands on science under changed conditions, demands that take contextualization into account, in principle without having to abandon the contrafactual demand for science's autonomy, independence, transparency, and public nature.

Bibliography

Abels G, Bora A (2004) Demokratische Technikbewertung. Bielefeld

Angell M (2000 a) Is Academic Medicine for Sale? New England Journal of Medicine 342 (20): 1516–1518

Angell M (2000 b) The Pharmaceutical Industry – To Whom is It Accountable? New England Journal of Medicine 342 (25): 1902–1904

Angell M, Relman A (2002) Patents, profits & American medicine: conflict of interest in the testing & marketing of new drugs. Daedalus 131: 102–111

Bundesverfassungsgericht, Entscheidung vom 26. Oktober 2004, (German Federal Constitutional Court, decision of Oct. 26, 2004) http://www.bverfg.de/entscheidungen/rs20041026_1bvr091100.html

Luhmann N (1965) Grundrechte als Institution. Berlin

Luhmann N (1990) Die Wissenschaft der Gesellschaft. Frankfurt

Nowotny H, Scott P, Gibbons M (2001) Re-Thinking Science. Cambridge

Nowotny H (2005) The Changing Nature of Public Science. In this volume: 1

Pestre D (2005) The Technosciences between Markets, Social Worries and the Political: How to Imagine a Better Future? In this volume: 29

Schmidt-Aßmann E (2005) Free Access to Research Findings and its Limitations. In this volume: 109

Schulze-Fielitz H (2005) Responses of the Legal Order to the Loss of Trust in Science. In this volume: 63

Stichweh R (2005) The Multiple Publics of Science: Inclusion and Popularization. Soziale Systeme 9: 210–220

Trute HH (2005) Democratizing Science: Expertise and Participation in Administrative Decision-Making. In this volume: 87

Trute HH (2005 a) Von der Ordinarien- über die Gruppen- zur Managementuniversität. Juristenzeitung 2005 (forthcoming)

Weingart P (2001) Die Stunde der Wahrheit. Weilerswist

Responses of the Legal Order to the Loss of Trust in Science

Helmuth Schulze-Fielitz

I. Posited question: Declining trust in science?

Among the insights of recent scientific research that enjoy a broad consensus is the view that the trust of the public in certain spheres of science has diminished, particularly in biomedicine and its technological application (cf. *Shapin* 1995: 392 ff., 398 ff.). Their moral legitimacy, it is argued, has dropped significantly. The consequences are that the professional autonomy of the sciences is restricted in their decisions about how to pursue scientific questions, and that the relationship of the sciences to political decision-making processes is undergoing a fundamental shift. Science's capacity to legitimate decisions made by political decision makers is becoming increasingly problematic (*Cozzens and Woodhouse* 1995: 533 ff.). Can outward manifestations of a crisis of confidence be substantiated empirically?

1. General reduction of trust in science?

It is doubtful that there is a general reduction in the public's trust in science (*Weingart* 1997: 606 f.). Public interest in science, especially in medical discoveries and technological inventions, remains as high as ever (cf. statistics listed in *Stehr* 2003: 314); the majority of changes in *technology* are regarded as positive (cf. *Noelle-Neumann and Köcher* 2002: 859). Though only 50.4 % of the Europeans surveyed in the Eurobarometer of the year 2001 affirmed that the benefits of *science* are generally greater than its detrimental consequences (as opposed to 61.2 % in 1992, cf. European Commission, Generaldirektion Forschung 2001: 25 f.), is hardly clear that this can be interpreted as a general decrease of trust in science; after all, the feeling that the progress of science continues to make life simpler and simpler is increasing (for data on Germany, cf. *Noelle-Neumann/Köcher* 2002: 858).

2. Areas of science

Consulting survey responses to this question, for instance from Great Britain, reveals a number of more differentiated findings (cf. The United Kingdom

Parliament 2000: Appendix 6). They reveal that only *certain fields of scientific research* are regarded to be controversial, that is, primarily the biosciences (biomedicine, biology), but also fields like the environment, reproductive medicine, communications technology and protection of privacy. This is true even within the individual scientific fields: generally, bio*medical* research enjoys broad support by the general public (*Cozzens and Woodhouse* 1995: 538.); here strong reservations are expressed only for individual application prospects, many of which, like the cloning of humans, have been decisively rejected by the public. (Human) genetic tests and biomedical research are regarded as considerably more beneficial, less risky, morally less problematic and, all in all, more worthy of support than genetic changes in food production, experiments with animals or xenotransplantation (The United Kingdom Parliament 2000: Table 4). Accordingly, when the public is asked about its hopes for benefits through scientific developments, most respondents list first the area of medicine and medications (57 %), the transplantation of organs (51 %), the eradication of diseases (43 %) and new possibilities for operations (31 %), but next to none mention genetically altered animals, plants or foods (only 1–2 % each) (The United Kingdom Parliament 2000: Table 6; parallel *Noelle-Neumann and Köcher* 2002: 873, 876). The help, or more specifically, the benefit or damage for other humans, is the standard used most often to judge to rightness of biomedical research. The public perception of science is characterized by cost-benefit considerations.

3. Scientific institutions

Losses of trust are valid not only for certain fields of science, but also for *certain institutions*, especially when political or economic *partial* interests impel such institutions to drive certain scientific developments (The United Kingdom Parliament 2000: Rn. 2.36.) whose advantages for the public are not clearly evident. In Great Britain the public has the greatest trust in BSE statements made by scientists from universities (42 %), followed by scientists in the meat industry (26.7 %) and consumer organizations (18.0 %); only 4.6 % trust government scientists on this issue. This is hardly a problem restricted to the BSE issue; similar values emerge for statements on the safety of nuclear power plants (The United Kingdom Parliament 2000: Table 2).

With regard to the trustworthiness of those professions or institutions that make decisions in the *biosciences*, physicians enjoy the highest trust (69–71 %), followed by advisory bodies consisting of a variety of specialists (62–66 %), experts (54–56 %), environmental groups (56 %) and consumer organizations (48 %). Considerably less trust is associated with all other institutions, including patient and animal protection groups (35 %), the government (35 %), the church (22 %), and industry (11 %) (The United Kingdom Parliament 2000: Table 2). Despite specific national differences, this general tendency is also reflected in the responses to the Eurobarometer survey questions regarding science and scientists (European Commission, Generaldirektion Forschung 2001: 38, Table 26).

II. Explanations and problems posed

Such findings demand general explanations. Are they evidence of crises founded only in the development of the relevant specialized field (especially that of biomedicine), or are they problems that reflect general developments in the system of science? In the latter sense I would like to argue in support of the thesis that such crises of confidence are the logical consequence of a "democratization" of science, which is forced by the societal "contextualization" of science.

1. The societal contextualization of science

The discourse in the sociology of science contends that the present has witnessed a fundamental *change in the acquisition of scientific knowledge*, especially as regards the technical and natural sciences (cf. *Nowotny et al.* 2001, 2003: 179 ff.; earlier *Gibbons et al.* 1994). The claim is that the university has lost its central position in the production of knowledge to other institutions such as research centers, scientific institutions of the government, industrial research, and consulting firms, and that applied research has taken on increasing importance to the detriment of "pure" theoretical research. According to this view, scientific inquiry no longer occurs within the framework of conventional disciplinary borders, as its orientation toward application characteristically leads to more intensive transdisciplinary research in response to urgent practical problems. Scientific quality is thus no longer established through peer review within the sciences, but rather according to supplementary social, political or economic criteria with a view to its application. Finally, the argument runs, the production of knowledge is more intensively bound up in social accountability or responsibility, increasingly impelling scientists to orient their research toward social values. The contexts of development mentioned above evince a societal "contextualization" of science.

"Science" thus contextualized no longer covers just theoretical research in the strict sense, but also the *process* of the manufacture of scientific knowledge and its realization in the process of practical application; one indication is the increasing frequency of the phrase "science and technology." A distinction is made between the ideal types of a *sphere of primary scientific research* and a *transfer zone*, to which the results of primary research are conveyed to societal decision-making processes. Depending on the understanding of "science," such questions are regarded as either answerable by science, or as not answerable by science, but subject to political or legal decisions in a sphere of "trans-science" or science policy.

Knowledge is socially constructed in both spheres, but in different and differently immediate ways. The social construction of scientific reality initially referred to the construction of theory within science, whereas the new elements of social construction in the sense of "contextualization" appeared to relate more to applications. The sharp division previously practiced between scientific (theoretical) research, development of a product and application of a product increasingly gives way to the view that these are merely steps of one unified process. Even the first phases of research processes, problem definition and working out hypotheses and concepts, are subject to societal evaluation.

Contextualization is clearly visible in fields of science in whose discourse *ethical* arguments play an increasingly important role, as each succeeding wave of modernization appears to create a "higher moral need" (*Höffe* 1993: 14). This is especially true for newly developed areas of research in the life sciences and biomedicine, particularly in those phases at the beginning and end of human life which have been subject to more detailed investigation, but also wherever such methods as experiments with animals play a major role. The moral need is heightened further as the ideals of society become increasingly heterogeneous – as a consequence, new kinds of institutions such as ethics commissions are established to make it easier to deal with the diversification of such values (*Albers* 2003: 428 f.).

"*Economization*" means that the process of science is driven from the outset by the perspective of its economic exploitation (*Nowotny et al.* 2003: 182 f.; *Hoffmann and Neumann* 2003). This starts with financial support for theoretical research, whose supplementary task is to generate "innovations" with a view to future markets, e.g. the biosciences and nanotechnology (cf. *Cozzens and Woodhouse* 1995: 548 f.), and ends with structural reforms to institutions of higher learning in order to tailor them to the practical requirements of today's economy. Also connected with this is the fact that the time between academic research and industrial product application has been shortened so radically that it is hardly possible to speak of a downstream phase of specific application. Rather, today an intensive process of permanent feedback occurs (cf. *Weingart* 1997: 607, 610). At the same time the value of scientific research is closely linked with the perspective of rapid market exploitation.

Closely connected with economization is the attempt to influence the direction of scientific developments through *political* decisions that set priorities for the promotion of science, oriented at the same time toward the research capacity available and the many promising prospects in certain areas of science (*Nowotny et al.* 2003: 181 f.). These trends toward finalization are generally related to the broad objective of reinforcing the orientation of scientific research toward the future, with the goal of improving public health, social, economic and other conditions of human life in the medium term.

2. Loss of trust as a consequence of the democratization of science

If contextualization means binding science to social objectives that can be influenced and must be accounted for politically, then binding science to the political bodies of democratic government is merely a consistent conclusion. Many current development trends can be interpreted in the sense of an attempt toward the broadest kind of "binding fast" to democracy: responsibility toward society, accountability, an obligation to deliver one's debt to society, controlling, internal and external evaluation, competitive rankings, and improvement of the quality of training are expressions of the circumstance that the citizen of the state, as a taxpayer and financier of the science system, wants to know what happens with his or her money.

In this case "democratization" means (cf. *Weingart* 2003: 60 f.) integrating and taking seriously the lay perspective; making political-democratic decisions about the priorities of publicly financed research; providing access to scientific expertise in the sense of public availability; scientifically "legitimating" the rationality of political-democratic decisions, which are not supposed to violate established knowledge; and finally the selection of specialists to construct the knowledge needed to legitimate decisions. Such demands seem justified and inevitable in the framework of democracy.

Crises of confidence reflect the links that have developed between technical-scientific activity and the participation of the democratic public. Where the latter sees no benefit to the orientation of scientific developments, or believes that this benefit is outweighed by the associated risks, it remains guarded or even offers resistance. This is true no matter which scientific field is involved, although health risks and particularly high financial burdens are constantly in the foreground and even independent of the actual knowledge about the state of scientific development. A growing role in the field of the perception of science and its practical consequences has fallen to the increasingly important mass media, which bring problems to a head, bundle and amplify them (cf. *Powell and Leiss* 1997: 227 ff.). In keeping with their efforts to compete for attention, the media play advocate for their audience and dramatize scientific issues, slightly overstepping the limits of objective reporting (*Felt et al.* 1995: 275 ff.). Not even the humanities and social sciences can escape such pressure to justify themselves in the face of the general public.

3. Tasks of the legal order and jurisprudence

Legal rules contribute to the process of societal contextualization of the sciences, even in areas where new kinds of scientific conflicts have emerged and/or the trust of the general public in science has diminished or at least is threatened. The following thus concerns the *legal regulation of the societal contextualization* of sciences in the sense of the correct or optimum instrumental use of law to control organizations and decision processes. Ultimately this concerns creating security of expectations, a central foundation legitimating the modern state and its legal order. Often what is missing are material legal standards which could offer sufficient support for decisions on individual cases. Viewed from this perspective, the procedural rules on processes of reaching decisions gain in importance, especially under conditions of scientific uncertainty (*Scherzberg* 2002: 123, 126 ff.). It is also important to differentiate among the different types of scientific conflicts.

III. Typologizing scientific conflicts and their methods

The reduction of trust in science is associated with three ideal, quite dissimilar types of scientific conflicts whose legal codes of practice pursue diverse objectives and feature different kinds of argumentation.

1. Fundamental value conflicts

a) Singular characteristic: Inimical to compromise

The first type is *religious, political-philosophical or morally grounded value conflicts* in the research process or in the application of research results. Certain research orientations are generally rejected in the name of highly abstract superior moral or legal values, ideas or principles like human dignity, freedom, equality, solidarity, justice, responsibility, integrity, the Bible (e.g. research on human embryos, cloning). In such controversies no compromise exists that could satisfy all participants. For questions that cannot be decided rationally, especially in the case of diverging basic religious assumptions, the only possibility is to search for forms of "overlapping consensus" (*J. Rawls*) in the sense of equal coexistence for differing ethical (basic) convictions. Prerequisite to this is the acknowledgement of dissent among (equally entitled) alternative attitudes; and the procedure of seeking consensus required must take in the whole range of diverse moral opinions in society (*Kelly* 2003: 349 ff.; *Moreno* 1995).

b) Institutionalized forms: Ethics councils

Institutions like ethics commissions and ethics councils have a filter function in their search for answers to the challenges presented by new discoveries in the biosciences (*Fuchs* 2001; Committee 1995: 89 ff., 115 ff.). Common to such institutions is that what used to be authoritative in professional practice, the individual decision of a single scientist, has been replaced by the institutionalization of participation by outside parties in these decisions.

Legally it is possible to determine after the fact whether *legally binding decisions* have been made. Such decisions may be valid only for *single cases*; often these are based on affirmative assessments by ethics commissions, which are assigned the task of protecting patients and others from dangerous or surprising research by weighing out the prospective benefits against the prospective risks before clinical testing of a medication or medical product may begin. Once clinical testing plans have been approved by an ethics commission, they have cleared the hurdles of any legal objections (cf. *Rupp* 1999: 846 ff.), the researcher is relieved of his or her individual responsibility, and trust in the integrity of the research is restored (*Albers* 2003: 421).

In some cases *political or ethical recommendations for action* are made by privileged advisory bodies, which can bring to bear considerable binding effects or presumptive effects (e.g. ethics councils, advisory commissions). In this respect they are *abstract-general recommendations* for certain problem areas of science (e.g. recommendations of Germany's National Ethics Council on Embryonic Stem Cell Research). Whether these recommendations are observed, however, is at the liberty of external decision makers such as legislative bodies. Frequently such general recommendations are not observed; they serve to construct public opinion in the discourse on fundamental scientific issues (*Albers* 2003: 423 f.).

Legally non-binding ethical recommendations by ethics commissions are also possible *in individual* cases, however. Compliance with these is left up to the individual researcher (*Rupp* 1999: 841, 851 ff.; on the U.S., *Heitman* 1995: 418) and do not relieve the researcher of his or her responsibility (*Albers* 2003: 422). In practice, such recommendations primarily serve as legal judgments aimed to establish recognized rules for the medical arts (*Sommermann* 2003: 82; diff. *Albers* 2003: 426 ff.). Not even the judgment that an embryonic stem cell research project is "high ranking" and "lacks an alternative" is a purely ethical question, as it concerns the interpretation of legal concepts, albeit initially by experts outside the legal profession. All in all, decisions important for the legal order are being expanded by an ethical perspective in order to increase the law's efficiency and ability to learn.

c) Example: The debate on stem cell research

One example of a fundamental value conflict is the debate about "consumptive" embryonic research to create human embryonic stem cells. This case concerns a potential medical utilization of the special capabilities of embryonic stem cells. Their extraction entails the destruction of embryos, which many hold to constitute human life. While the production of embryos for research purposes is prohibited in Germany, in-vitro fertilization is not, which actually creates surplus embryos on a regular basis. Also permitted is research on stem cells already obtained through the destruction of embryos; stem cells may not be extracted, but can be imported as long as they were extracted abroad before 1/1/2002 and then preserved cryogenically.

Between the view that holds every human life conceived at the moment of fusion between sperm and egg to be equal to every human baby born (and thus absolutely inviolable), and those who regard human life and its dignity to be worthy of protection only once a baby has been born, the predominant reading seeks on the one hand to provide moral and legal protection for unborn life, while on the other hand weighing this protection against interests that run counter to it, often in the majority, that foster hopes for medical progress. The religious or philosophical bases of these opposing views remain fundamentally incompatible.

2. Methods of risk assessment

A second type involves conflicts about how to construct correct knowledge about *health or other risks* that can result from the technical implementation of new scientific findings. The point here is how to embed technical-natural scientific knowledge in the social and economic circumstances of citizens' lives: e.g. atomic power, genetically altered animals and plants, carcinogenic components of foods or industrial products, mobile telephone technology. Risk assessment methods should be applied to evaluate the dangers and risks the implementation of such knowledge poses for humans, and then these results weighed against the anticipated benefits.

a) Singular characteristic: Risk as a social problem

The first problem is that the level of risks the general public is willing to tolerate can be *set at very different levels* depending on the kind of risks and their *perception* by the public. The risks about which an individual can decide individually (for instance, driving an automobile) are accepted much more readily than risks that an individual cannot easily avoid (e.g. the food supply). The willingness to submit to risks is dependent on a profusion of additional variables: the extent to which these risks have been discussed with the public; additional benefits of acceptance; whether risks are uncontrollable, incalculable, inequitable, irreversible, affect all generations, or are of a potentially catastrophic scale, etc. The question of how safe is "safe enough" can never be answered by science alone.

A further problem is that the share of empirically certain knowledge in methods of risk assessment has to be supplemented by a high share of non-knowledge and presumptions about plausibility. Scientific specialists make political value judgments when they present such presumptions or forecasts and thus are to some degree involved in the political conflict (unless they reject an evaluation as "unscientific"). The *share of the socially constructed* in such *risk forecasts* is comparatively *quite high* (cf. *Jasanoff* 1990: 68, 75 f., 1987: 209 ff.; for an overview, *Pinch* 1981: 131 ff.). Scientific uncertainty and the necessity of balancing out incommensurable factors constitute a microcosmos of conflicting academic approaches, regulatory philosophies, national traditions, social values and modes of professional behaviour (*Cozzens and Woodhouse* 1995: 542; *Majone* 1984: 15 ff.; in detail, *Majone* 1989). In this respect no expert opinions can be refuted; they merely can be supplemented by expert opinions offering alternative evaluations. Therefore the personal integrity (including the independence and quality) of the specialists becomes a subject of public debate.

b) Institutionalized forms: Medical and technical-natural science advisory bodies

Most of the expert *bodies governing environmental and technology law* are executive-appointed panels of medical or technical-natural science specialists. They set standards (e.g. thresholds) and procedures to restrict the dangers and risks posed in their practical application by the administration. Danger is regarded as the "product" of the probability of a hazard occurring and the extent of the damage it would cause (should the events run their course unchecked), and is still viewed as an established dogmatic judgment, rooted in the understanding of the classical nineteenth-century laws restricting freedom in order to avert danger. The newfangled concept of risk goes beyond this to designate a state of affairs in which the prognosis of damage is characterized by a larger insecurity. This uncertainty concedes broad latitude as to how a risk may be determined as legally relevant, for additional degrees of probability are not established on a regular basis. It almost always permits the environmental or health policy demand that even "more" precautions be taken, for instance the reduction of thresholds.

c) Example: The debate on mobile telephone technology

Typical elements can be established in the *example* of the current German debate about the environmental impact of *mobile telephone technology* that suggests a crisis of public confidence in science (*Büllingen* 2003: 35 ff.). One symptom is that newspaper reporting about the impact of electromagnetic fields on the environment is increasing absolutely; the share of negative reports among these has risen to 82 % (2001), and most of these report on potential health risks. Despite their sensitizing effect, the potential health risks posed by mobile telephone technology appear to be estimated as relatively low: only 11.9 % held them to be "very high," significantly fewer than in the cases of asbestos (56.6 %), atomic power (45.6 %), traffic (36.5 %) or genetic technology (26.4 %); on the other hand, only around 5 % presume no risks at all. Eighty percent of those surveyed believe that it would make sense to reduce thresholds.

This debate about reducing thresholds is well underway outside of Germany, especially in those countries where introducing precautionary values lower than the internationally recommended thresholds has stimulated public discussion and led to demands for a downward "threshold spiral." One factor that plays a role here is that while the *thermal* effects of high-frequency electromagnetic fields have been studied in depth and are relatively uncontroversial, their *non-thermal* effects have not; from this perspective discussions focus on such possible consequences as cancer illnesses. It is not proven that high-frequency fields can cause cancer, but this possibility cannot be ruled out for lack of precise scientific findings (*Scherer and Schimanek* 2003: 299 ff., 313). The valid thresholds set by regulatory authorities are oriented only on the thermal effects of electromagnetic fields; non-thermal effects are not included. The thresholds ultimately are based on internationally recognized recommendations and guarantee protection from known health risks and significant disturbances, and appear to correspond with the state of scientific knowledge. No objections can be raised about their constitutional legality until it becomes recognizable that these threshold values are fully inadequate to protect human health; the state does not need to take precautions against purely hypothetical threats to human health (Bundesverfassungsgericht 2002: 1639).

3. Methods of balancing interests

a) Singular characteristic: Building social compromises

A third type of scientific confrontation is connected with questions of political, social and economic development and application in which scientific expertise may be integrated, but is so closely linked with *the process of defining societal-political value* that scientists participate in decisions about central questions of valuation. In contrast to the cases in which conflicts about values or risk arise, here it is even more difficult to scientifically separate definite knowledge from prescriptive conclusions. Such scientific conflicts occur outside the area of techno-science, essentially in the mantle of *conflicts in (social) policy design*. The *dimension of interest* is of primary importance in such application-oriented conflicts. They regularly conceal central questions of justice for which there are no clear scientific answers. Therefore such confrontations can function as more than mere discourses on justice – grounded in social philosophy – to expose diverging social interests: In most cases it is possible to find compromises on the level of opposing interests, which is generally out of the question in conflicts about fundamental values(evaluation). Whenever interests can be translated into quantities of such variables as money or time, compromise becomes feasible.

At first glance the form of scientific conflict concerned here would not appear to be new, but located on the same logical level as the new ethical problems of the life sciences and the stipulation of quantities to serve as danger thresholds to protect from unknown risks. However, at second glance a new quality of these forms of (social) scientific knowledge emerges from the growing orientation of contextualized science toward the long-term future. This quality also has an impact on the question of trust in science. Increasing knowledge (and at the same time, ignorance) and/or the increased complexity of political decisions, as a consequence of lengthening causal functional chains, raise the demand for scientific input in political decisions with long-term consequences (e.g. the level of pensions in 30 years). Forecasts supported by science become dependent on a greater and greater number of variables, each of which has a different impact on the results; the many possibilities of socially constructing ceteris-paribus assumptions and other conditions make recommendations that can be utilized in practice dependent on non-scientific assumptions. The difference between an established, empirical-analytical (social) scientific expert opinion and assumptions based on less well-founded scientific knowledge, but transformed into plausible "socially robust knowledge" (cf. *Nowotny et al.* 2003: 191 f.), is becoming just as blurred as the difference between expert opinions and prescriptive political directives for action.

b) Institutionalized forms: Social science advisory bodies

Governing bodies in which *sociological* or special theoretical economic *expertise* are *linked with* the *imperative of action by the political system* can be understood as institutions that filter scientific expert opinions on this third level of scientific conflict. They generally take the institutionalized form of councils of specialists functioning as advisory organs, permanently or temporarily, primarily for the federal government. Such councils usually combine representatives of the four areas of science, the organized public, private industry and politics or administration (*Siefken* 2003: 484). These bodies elucidate matters of concern, reinforce the positions of the government, and, if need be, solve problems symbolically and/or negotiate a consensus (e.g. the "Rürup Commission" on pension reform).

c) Example: The debate on pension reform

One *example* is the various (Europe-wide) conflicts about reforming *old-age security systems*. Their ability to function presupposes an enduring trust in the institutions on the part of the insured, who must believe that the system can reliably guarantee the payments to which they have a claim for 50 years and longer (*Hinrichs* 2000: 283). Changes in regulations require extremely long-term political decisions about how to shape the system, whose necessity and scale are dependent, in turn, on quite diverse assumptions that can lead to very different long-term shifts in the forecast. They thus provide great latitude for social construction in the choice of forecasts and the resolution of problems through compromises. As a consequence, the social and economic advisors involved are linked very closely with normative assumptions about how to shape society and thus with the area of political consulting. This in turn makes them subject to the logic of balancing out various interests, which generally can be solved by compromises according to financial compensation.

For instance, everyone involved in putting together the pension reform passed in Germany in 1989 believed that additional reforms would be superfluous for the next 20 years. Between the alternatives of sparing pensioners from reductions in payments (by doubling the level of contributions 40 years later) and fixing the level of contributions (by halving the level of payments in the long term), a medium-term perspective extending beyond the next few legislative periods led to a system-immanent compromise designed to maintain the stability of trust in the system. Nevertheless, as typical, the future recipients of payments bear the main burden of the reform, and the current taxpayers and contributors the lesser part. The social security system's "blessing of opacity" facilitated the institution of such changes, which remain incomprehensible for the individual and are little more than a series of "adjusting screws" that can be calculated into financial quantities (such as the level of taxpayer subsidy, the percentage of salary to be paid in contributions, the age of retirement) for the social system of pension insurance.

4. Conclusions from typological differences

The above typology makes clear the characteristic differences between the various kinds of scientific conflicts – and the practical consequences they entail. Under certain criteria these can be made clear through cross-sectional comparisons, such as the degree of amalgamation of science and politics, or the intensity of obligations to justify scientific work, to name just a couple. Just two examples are outlined below: the differing importance of majority rule and appointing people to scientific bodies.

a) Problems in shaping majorities and compromises

A characteristic feature of consulting or decision-making bodies is that they consist of multiple individuals collectively. As in other such bodies, this means that the principle of the majority is the predominant rule for making decisions. In principle, it appears that there is nothing to be said against the fact that disputed scientific issues can also be decided through majority voting – be it in ethics commissions or in scientific councils. But can values conflicts be resolved by decisions in cases of problems (like, for instance, the status of the human embryo), for which no agreement on the national or international levels (*Fuchs* 2001: 73, 88 ff., 94 f., 95 ff., 102 ff., 120 ff.) can be reached – and where the best hope is to structure dissent? Where compromises are sought, the controversial points of departure must be backed by interests that can be graduated step-by-step and balanced out against each other to find a compromise. This is not only (typically) possible for processes of balancing interests, but also for determining the acceptability of social and technical risks.

Value conflicts such as those in the bioethics discourse generally cannot be translated into interests that can be negotiated, however – in this respect there are also limits to bioethical decision-making according to majority decisions. Thus we are faced repeatedly with strong tendencies to withdraw certain questions from majority decision-making because they are not accessible to this method. In German discussions about bioethics this is achieved mainly by qualifying certain kinds of research as violations against human dignity as protected in Article 1, paragraph 1 of the German Basic Law; this makes them clearly inadmissible and withdraws them from any consideration, compromise shaping or majority decision. What is more, majority rule measures the number of the votes only quantitatively, leaving out such factors as the religiously-founded intensity of the position behind a vote (*Heun* 1983: 209 ff.). Especially in polarized values debates, a clear majority view held by indifferent, superficially founded "Yes" votes may be countered by an existentially affected minority of a few opponents whose intensity in fact annuls the formal majority. For bodies relevant to bioethics, this can also lead to a situation in which majority decisions (in some cases with minority votes) can not be made at all, but instead different views are (re)presented next to each other without any further qualification (e.g. in the German National Ethics Council, cf. Nationaler Ethikrat 2003: 74 ff., 106 ff.). The majority rule obviously has a different status in value conflicts than in other scientific conflicts.

b) Problems in putting together scientific advisory bodies

A characteristic of the various bodies in scientific conflicts is their often quite heterogeneous composition of individuals, pluralized in terms of expertise, politics, political parties and professionals. How does this affect the various scientific conflicts? In this respect it appears important whether decisions are made on the criterion of the *specialized scientific* profession (e.g. ethics councils at schools of medicine, scientific ethics councils of state ministries) or whether the professional expertise is linked with politicians in order to prepare further *political decisions* (e.g. Enquete commissions of the Bundestag; to a lesser extent also the National Ethics Council) or – as for example, in parliaments – are decisively influenced by general political considerations, or whether the affected are also *included* in the consultations according to the model of participatory assessment of the technical consequences. Depending on the kind of scientific conflict, the composition of decision-making bodies may be oriented more toward societal pluralism or more toward the specialized members of the profession.

The following text (V.) hence concentrates on questions associated with the organization and procedures of bodies that deal with value conflicts. These, too, are subject to certain conditions of constitutional law (IV.).

IV. Constitutional-law aspects informing legal conclusions

When recent developments are viewed in a larger perspective, it is possible to recognize "guiding aspects" (grounded in constitutional law), which are shared by all institutionalized contexts and decisive for the responses of the legal order to the increase in scientific conflicts. This is true in principle for all kinds of scientific conflicts.

1. The core question behind the debate often concerns the *limits of entrepreneurial and/or scientific freedom* in the liberal democracies of the "Western constitutional state" variety. Due to the German Basic Law's fundamental

presumption of the right to make use of constitutionally protected freedoms, all scientific conflicts prove to be conflicts about the restrictions of freedom, above all scientific freedom or professional freedom; in the framework of the distributional conflicts faced by the welfare state, they are also about the freedom of property and general freedom of action. All of these conflicts concern the reasonable criteria under which the legislator can stipulate concrete restrictions to public welfare. These decisions have such explosive potential because the presumption of freedom in the Basic Law is restricted according to ethical aspects, whose rigor in terms of constitutional law is doubtful.

2. A further aspect shared by these conflicts, despite all tendencies toward contextualization, is the importance of *protecting science's epistemological determination by inner laws* in the social transfer process through quality assurance of scientific advice. The independence from political conditions of the scientific consultation process in the narrow sense is an important prerequisite for the quality of scientific advice (*Cozzens and Woodhouse* 1995: 549 f.).

3. *Pluralism* is also a central normative aspect for the qualified resolution of scientific conflicts. A multiplicity of various scientific approaches, types of specialized knowledge and institutional experiences can extend the spectrum of aspects to be deliberated. Filling advisory councils with as much diversity as possible should generate factual conflicts and their discussion, as well as reasonable results in the construction of a scientific reality that can react responsively to the multiplicity of social interests.

4. Finally, *transparency, representativeness, influence* and the *earliness and punctuality* of the participation of the public in decisions about scientific conflicts are also becoming central criteria for judging the evaluation of science. In the "transfer zone" that links scientific and non-scientific expertise in the form of boundary organizations, non-specialists (laymen, stakeholders, politicians, scientists in related fields) are integrated to help make decisions about fundamental scientific-strategic issues, because fundamental decisions about research are made less autonomously by the scientific system than they used to be.

V. Consequences for the practice of law: The special example of ethics councils

1. Controlling the context of science through the legal order

There has long been a broad consensus within jurisprudence that certain specialized areas largely elude any kind of legal regulation through imperative commands. Actors in these areas may be subject to societal regulations, but are for the most part free to make autonomous decisions. Because the effectiveness of a legal system is contingent on its acceptance by societal actors (e.g. motivation for voting, taking initiative, innovation and involvement on a superindividual level), individuals who value their special sway over exchange and prevention more than these advantages can avoid state control. In this case legal order is reduced to an indirect control function and can regulate only the conditions that encourage certain behavior in society. Such "contextual control" is largely the case in the field of science as well. The existing political and legal rules and options relevant for such contextual control are discussed below.

2. Structuring advisory bodies according to the legal order

The way in which scientific advisory or decision-making bodies are structured in terms of organizational or procedural law can have considerable consequences for the material results of their advice. A clearly defined task alone is not sufficient. With the variation in task descriptions and procedural rules, combinations of personnel and obligations of the participants (e.g. registration and documentation, submission of reports, participation and authorization, cost regulations, etc.), one and the same material mission can be realized in extremely different ways. Significant differences also follow from whether legal or ethical guidelines are prescribed or first must be found, whether the matter involves decisions for the general or the individual case, whether factual or legal veto positions exist, how many people must be included in the bodies in order to fulfil a task within a prescribed deadline, how "representative" or one-sided their members and according to what criteria they were selected, to which hierarchical level they are assigned within an organization (e.g. a clinic) and who is responsible for them (i.e., the doctors themselves or the administration) and much more. Despite the critical importance of organizational rules, they are not legally stipulated for the great majority of such bodies. Occasionally legal regulations regarding the establishment and membership of ethics commissions are demanded (*Gramm* 2001: 625; *Rupp* 1999: 843 f.), especially when expectations about the practical or legal impact of their deliberations (and thus the character of their intervention) are particularly high.

a) Organization: Constitution and composition of members

Bioethical decision making bodies also exhibit different typologies (*Gross* 1999: 51 ff.). For instance, they can be attached to *parliament* or to the *government and administration,* or distinguished by their *kinds of membership* (honorary, full-time or part-time; long-term or short-term), whereby in bodies relevant for bioethics the number of voluntary and part-time offices clearly exceeds the number of full-time members: The sensitization to bioethics acquired by practicing another kind of full-time activity is considered a benefit in this field.

One can further differentiate in terms of *membership mechanisms* (election, nomination or delegation). The composition of the bodies and the results of their deliberations can be (partially) directed through such methods as a one-sided nomination policy; in contrast, electing councils or entitling a plurality of organizations to send delegates can safeguard against bias. A systematic response to the problem of societal pluralities is imperative for commissions created as permanent bodies, e.g. the distribution of rights to nominate members; the stipulation of specific qualifications for members; rotation duties based on temporally restricted terms of office; quotas, and the like. All of the above should balance out the various scientific cultures or religious "families of thought." In any case it appears quite problematic when ministers are able to appoint people to scientific advisory bodies of the federal government according to their own scientific preferences alone (cf. concrete examples for Germany at *Siefken* 2003: 499).

Another important point is whether these bodies are pluralistic or tailored to certain professions (*Gross* 1999: 61 f., 63 ff., 87 ff.). In *pluralistic bodies* citizens or affected parties bring their multiplicity of interests and views into the cooperative deliberations; the members of *professional bodies* are selected according to their specialized competence in order to exercise their authority on the subject at hand in the deliberations of the body. Many bioethics councils appear at first glance to be of the more professional type, e.g. ethics commissions, in the stricter sense of medical control through colleagues (peer review) or in the broader sense of higher-level control through outside parties, such as lawyers, moral philosophers or theologians (community review).

At second glance, their members can be selected at least according to pluralistic aspects *as well*. This is true for the participation of affected parties (e.g. patient representatives) and nursing staff in ethics commissions, and generally becomes all the more true the more closely such bodies are associated with political consulting in the broader sense. The rights to nominate specialized members to advisory commissions of the German Bundestag are subject to the standard proportional representation by parties; thus the associated political pressure to conform also affects the specialists. Different members of the National Ethics Council established by the federal government are also closely associated with certain societal interests (e.g. churches, associations of the disabled). The more "politically" bioethics councils are composed, the more likely they will be oriented toward the logic of compromises and majority decisions. The more the bioethical recommendations of such bodies are directed toward abstract, general questions of political development that transcend individual cases, the closer they are to politics.

b) Method: Making decisions in the space between autonomy and heteronomy

Procedural rules influence the contents of decisions and guarantee their implementation. This begins with the way *the tasks and topics are formulated.* A frequent concern is the "ethical tenability" of certain scientific projects. Ethics commissions can only take action in response to a motion by an authorized party and must cease their activities when the motion is withdrawn, although such instances are quite rare due to the de facto compulsion to request the opinion of an ethics commission. Are the tasks of the commission predetermined or does the body have the possibility of determining the object of deliberations itself? Obligating an applicant to present a scientifically founded explanation in accordance with the stem cell law, for instance, or granting an ethics commission only the authority to investigate submitted materials, prevents independent determination of facts of a case and precludes any scientific judgments extending beyond a simple plausibility check; written circulation procedures appear to predominate.

Participation by outside parties can influence decisions; even simple duties to document, report or substantiate decisions can have enduring consequences on the decision-making and implementation processes. Positive evaluations (in contrast to rejections or reservations) need not be explained by ethics commissions; however, a general *duty to substantiate decisions* would bring with it a significant workload, especially since it appears that ethics commissions seldom discourage a researcher from realizing a project, but often raise objections to details. Advisory bodies closely associated with politics never conduct regular public deliberations (with the recent exception of the new National Ethics Council), but always take the *public* as their central point of reference. Local ethics commissions' decisions about individual cases are often not made public at all. As a tendency, public procedures and access to information should gain in importance for such advisory bodies as well (*Albers* 2003: 435).

Special ethics commissions on medical consultation cannot replace the permanent responsibility of physicians to advise patients, but can serve as a judgment to protect such consultations from attack and thus allow a doctor to take on more responsibilities. Even the difference between the practices of granting the project director a *personal hearing* versus requesting written testimony can have varying results. In either case the *intensity of investigation* of research projects is limited.

c) Limits on regulating highly personal value judgements

The decisions made by bioethics councils are characteristically the summation of the highly personal value judgments of the individual members of the body; even if a general substantiation duty were to be instituted, such values can be structured legally only to a limited degree. The legal dogma of administrative court control in Germany is cognizant of administrative decisions (e.g. state investigations) for which the administration is granted "latitude of judgment", but whose lawfulness can be verified by the court only to a limited degree: namely for correctly identifying the facts of the case, compliance with procedural law, correct interpretation of the law, and the safeguarding of universally valid bases of valuation, as well as by identifying any influence exerted by considerations hostile to the case, such as false assumptions about its factual content. On the merits of the case, the correctness of a decision cannot be decided in court.

The same is true for the attempt to determine ethical positions in the framework of or in accordance with legal prescriptions: The correctness of decisions made by cooperative councils on bioethics can be influenced only by means of their organizational structures and methods, but cannot actually be ensured according to predetermined standards with any precision (when, for instance, are the objectives of stem cell research "high ranking" in the sense of the law?). To this extent an ethics commission can only reject a research plan by identifying it as evidently untenable. Therefore cooperative bodies with majority decisions, as opposed to single monocratic officials, are the appropriate authorities for making decisions or providing advice in this area. Members of ethics commissions are doctors, often joined by philosophers, theologians, legal scholars, social scientists and even other types of natural scientists. All are unified in a common goal: to discover bioethical standards of behavior, to work out the pros and cons of these standards and to justify them (*Taupitz* 2003: 818); sometimes they also aspire to define these standards morally or legally. All members must transcend their professional standards based on specialized expertise (*Kelly* 2003: 344, 350 f.; *Moreno* 1995: 66). The decisive value decisions about bioethics, especially in the controversial areas, are based on such foundations as religious faith, a feeling of justice, the ability to discriminate, ethos or even the simple everyday intuition of participating council members. Here (e.g. in the National Ethics Council) the ethical value judgment of a bishop is institutionally equal and of equal weight to that of a representative from an interest group or of a Nobel prize-winning natural scientist: The same bioethical aspirations make everyone equal.

The members of such councils deliberate as proxies for the general public sphere of citizens; the integration of non-specialized expertise and in some cases even of laymen is an expression of trust in the broad distribution of the discriminatory power throughout society (*Fuchs* 2001: 16; *Jennings* 1991: 461). The more a scientific conflict is carried out in the public and with the participation of laymen, the more intensely it can deviate from traditional scientific standards and find those tactical-strategic arguments and forms of strategies of influence that characterize the process of political confrontation as well. Thus the task of legally optimization scientific advising is to create conditions for the preservation of scientific standards of rationality such that scientific arguments are not neglected in favor of social, political or other interests.

3. Legally ensuring the quality of advice

The task of *ensuring the quality of scientific advice* and of guaranteeing sound knowledge has also been recognized by the European Commission and formulated in guidelines for all departments of the commission (cf. European Commission 2002). The (hollow?) catchwords are: accountability, pluralism, integrity, openness, participation, effectiveness, coherence, commensurability and subsidiarity. These have been condensed into three "core principles": the excellence, openness and effectiveness of scientific advising. Three factors are supposed to be decisive for quality in particular: subject expertise, acting in an independent manner and pluralism. With a view to ethics councils I would like to address only the criteria of pluralism and of transparency and publicity.

a) Pluralism

The aspect of *pluralistic diversity* is also an important aspect for the quality of scientific advising, reflected on the highest levels in the stipulation that at least 40 % of the members of both sexes be represented in advisory bodies (European Commission 2002: Ziff. 4.8.); on the European level this is extended by the ideal of multinationality. To this end scientific bodies are filled according to political or pluralistic criteria. This does not eliminate the boundaries between science and politics; rather, the scientific standards of rationality remain in the councils (*van den Daele* 1996: 319). However, when scientists' answers overstep their established knowledge by responding with political or other value decisions, they face (the possibility of) critique. Even if no consensus can be reached, such criticism yields learning processes and exchange in the form of debate (*van den Daele* 1996: 321, 323 ff.; *Majone* 1989).

A central task in constituting professional and/or pluralist bodies is therefore guaranteeing their representativeness while taking adequate consideration of the views of minorities and unconventional streams of thought. This intentional balancing of diverging interests or inclusion of "counter-experts" is often the way the neutrality of advisory bodies is ensured. With this, certain minority positions or interests whose arguments are easily neglected by elites in the process of making decisions, such as consumer interests, can be *upheld through legal regulations* (cf. *Cozzens and Woodhouse* 1995: 550). For the ethics commissions of the medical associations in Germany, such demands for pluralism are largely left up to the local medical associations themselves. An investigation specific to the specialized area is required to determine whether the institution of a legal framework is a suitable way of ensuring that demands for pluralism are met. One variant is *indirect* pluralization: The ordinance of the European Authority for Food Safety, for instance, does not include any rules for the composition of the various scientific bodies, but stipulates the makeup of the administrative council that selects and nominates the members of these bodies: Of 14 members of the administrative council, four must come from within organizations "that represent the community of consumers and other interests in the food chain."

b) Transparency and publicity

Transparency is an additional central requirement that allows the public to recognize and monitor the measures taken to ensure the quality of scientific advice. Transparency not only with regard to the choice of topics, but also to the way consulting tasks are allocated to specialists and the way their results are assimilated into the decision process, makes it possible to mobilize citizens for quality assurance. *Procedural transparency* thus is a central prerequisite that also ensures the quality of advising. In the case of ethics commissions, such demands must be applied in a modified form to take account of such factors as patients' protection. At any rate changes to procedural law, such as those in the National Ethics Council, are made public on a regular basis, and after initial reservations, this body's meeting are now open to the public.

Science lives from *publicity* and public critique. Science directed by the state administration participates in its need for confidentiality and thus opens a non-public field of secretiveness and scientific suspicions that appears suspect, especially since latitude for subjective evaluation always remains, even for those who are ignorant about certain issues. Confidentiality in the administration of science reduces the public's trust in both the administration and its scientifically guided conclusions. To guarantee the highest degree of transparency, it will be necessary to make accessible to the public not only all expert opinions worked out for government and administration by outside parties, but also all responses to these by members of the government and administration. In this case scientific expertise must also make clear where uncertainty remains and refrain from trying to hide it – for this is what generates suspicion rather than facilitating trust. Conversely, the body must fulfil its reporting duties in order to make objective, reliable and easily accessible information available to the public and to those affected. This is also true for ethics councils to the extent that their consultations or recommendations can have an influence on the general or specialized public (Committee 1995: 179).

VI. Outlook: The task of a science law

The above considerations present building blocks for an area of law that has yet to develop: *science law*. No matter how much it is influenced in this practice indirectly by legal rules, only the scientific process itself in the narrow sense can decide about the truth or correctness of scientific research; in this sphere law can develop only limited power of jurisdiction. Possible, and perhaps required in view of their actual importance for the scientific process described, are legal rules to control the social context of scientific research, in order to optimize science as a form of life, a means of life and a task for the future. This also includes the regulation of scientific conflicts. In this respect science law will become a long-term mission.

Bibliography

Albers M (2003) Die Institutionalisierung von Ethik-Kommissionen. Kritische Vierteljahresschrift für Gesetzgebung 86: 419–436

Büllingen F (2003) "Elektrosmog" durch Mobilfunk? Akzeptanz und Risiko im Licht der öffentlichen Debatte. Aus Politik und Zeitgeschichte 42: 35–46

Bundesverfassungsgericht (3. Kammer des Ersten Senats). Neue Juristische Wochenschrift (NJW) 2002: 1638–1640

Committee of the Social and Ethic Impacts of Developments in Biomedicine, in: Bulger RE, Bobby EM, Fineberg HV (eds) Society's Choices. Social and Ethical Decision Making in Biomedicine, Washington D.C. 1995, pp 27–194

Cozzens SE, Woodhouse EJ (1995) Science, Government, and the Politics of Knowledge, in: Jasanoff S, Markle GE, Petersen JC, Pinch T (eds) Handbook of Science and Technology Studies, Thousand Oaks, pp 533–553

van den Daele W (1996) Objektives Wissen als politische Ressource: Experten und Gegenexperten im Diskurs. In: idem, Neidhardt F (eds) Kommunikation und Entscheidung. Politische Funktionen öffentlicher Meinungsbildung und diskursiver Verfahren, Berlin, pp 297–326

European Commission, Generaldirektion Forschung (ed) Eurobarometer 55.2: Wissenschaft und Technik im Bewusstsein der Europäer, December 2001

European Commission (ed) Commission communication on the collection and use of expertise by the Commission: Principles and guidelines, COM (2002) 713

Felt U, Nowotny H, Taschwer K (1995) Wissenschaftsforschung. Eine Einführung. Frankfurt

Fuchs M (2001) Internationaler Überblick zu Verfahren der Entscheidungsfindung bei ethischem Dissens. Gutachten für die Enquete-Kommission "Recht und Ethik der modernen Medizin" des 14. Deutschen Bundestages

Gibbons M, Limoges C, Nowotny H, Schwartzman S, Scott P, Trow M (1994) The New Production of Knowledge. London

Gramm C (2001) Verrechtlichung von Ethik und Ethisierung des Rechts. In: Bohnert J, idem, Kindhäuser U, Lege J, Rinken A, Robbers G (eds) Verfassung – Philosophie – Kirche. Festschrift für Alexander Hollerbach zum 70. Geburtstag, pp 611–625

Gross T (1999) Das Kollegialprinzip in der Verwaltungsorganisation. Tübingen

Heitman E (1995) Institutional Ethics Committees: Local Perspectives on Ethical Issues in Medicine in: Bulger RE, Bobby EM, Fineberg HV (eds) Society's Choices. Social and Ethical Decision Making in Biomedicine, Washington, D.C., pp 409–431

Heun W (1983) Das Mehrheitsprinzip in der Demokratie. Berlin

Hinrichs K (2000) Auf dem Weg zur Alterssicherungspolitik – Reformperspektiven in der gesetzlichen Rentenversicherung. In: Leibfried S, Wagschal U (eds) Der deutsche Sozialstaat: Bilanzen – Reformen – Perspektiven, pp 276–305

Höffe O (1993) Moral als Preis der Moderne. Ein Versuch über Wissenschaft, Technik und Umwelt. Frankfurt

Hoffmann D, Neumann K (eds) (2003) Ökonomisierung der Wissenschaft. Weinheim

Jasanoff S (1987) Contested Boundaries in Policy-Relevant Science. Social Studies of Science 17: 195–230

Jasanoff S (1990) American Exceptionalism and the Political Acknowledgement of Risk. Daedalus 119: 61–81

Jennings B (1991) Possibilities of Consensus: Toward Democratic Moral Discourse. The Journal of Medicine and Philosophy 16: 447–463

Kelly SE (2003) Public Bioethics and Publics: Consensus, Boundaries and Participation in Biomedical Science Policy. Science, Technology, & Human Values 28: 339–364

Losch B, Radau C (2003) Forschungsverantwortung als Verfahrensaufgabe. Neue Zeitschrift für Verwaltungsrecht (NVwZ): 390–396

Majone G (1984) Science and trans-science in standard setting. Science, Technology, & Human Values 9: 15–22

Majone G (1989) Evidence, Argument and Persuasion in the Policy. New Haven

Moreno JD (1995) Deciding together: Bioethics and moral consensus. New York

Nationaler Ethikrat (2003) Genetische Diagnostik vor und während der Schwangerschaft. Berlin

Noelle-Neumann E, Köcher R (eds) (2002) Allensbacher Jahrbuch der Demoskopie 1998-2002. München

Nowotny H, Scott P, Gibbons M (2001) Re-Thinking Science. Cambridge

Nowotny H, Scott P, Gibbons M (2003) Introduction: "Mode 2" Revisited: The New Production of Knowledge, Minerva 41: 179–194

Pinch TJ (1981) The Sun-Set: The Presentation of Certainty in Scientific Life. Social Studies of Science 11: 131–158

Powell D, Leiss W (1997) Mad Cows and Mother's Milk: The Perils of Poor Risk Communication. Montreal et al.

Rupp HH (1999) Sind Ethik-Kommissionen Rechtsausschüsse und ihre Voten Verwaltungsakte? In: Kästner KH, Nörr KW, Schlaich K (eds) Festschrift für Martin Heckel, Tübingen, pp 839–856

Scherer J, Schimanek P (2002) Rechtsfragen elektromagnetischer Felder ("Elektrosmog"), in: Hendler R, Marburger P, Reinhardt M, Schröder M (eds) Jahrbuch des Umwelt- und Technikrechts 2002: 295–317

Scherzberg A (2002) Wissen, Nichtwissen und Ungewissheit im Recht. In: Engel C, Halfmann J, Schulte M (eds) Wissen – Nichtwissen – Unsicheres Wissen, pp 113–144

Shapin S (1995) Trust, Honesty and the Authority of Science. In: Bulger RE, Bobby EM Fineberg HV (eds) Society's Choices. Social and Ethical Decision Making in Biomedicine, Washington, pp 388–408

Siefken ST (2003) Expertengremien der Bundesregierung – Fakten, Fiktionen, Forschungsbedarf. Zeitschrift für Parlamentsfragen 34: 483–504

Sommermann KP (2003) Ethisierung des öffentlichen Diskurses und Verstaatlichung der Ethik. Archiv für Rechts- und Sozialphilosophie 89: 75–86

Stehr N (2003) Wissenspolitik. Frankfurt

The United Kingdom Parliament (ed) (2000) Select Committee on Science and Technology, Third Report

Weingart P (1997) From "Finalization" to "Mode 2": old wine in new bottles? Social Science Information 36: 591–613

Weingart P (2003) Experte ist jeder, alle sind Laien. Gegenworte 11: 58–61

Democratizing Science: Expertise and Participation in Administrative Decision-Making

Hans-Heinrich Trute

Introduction

In recent years, flying the banner "Democratizing Science", social studies of science as well as policy makers have advocated greater public participation in science, policy and legal decisions concerning science and scientific knowledge. I will pick up this claim for democratization of science, although very ambiguous in its exact meaning. It covers very different phenomena, the inclusion of lay people in research, participation of the public in scientific controversies, in political processes as well as administrative procedures concerning scientific matters, observation of science by media and the like (*Trute* 2005). Although a kind of social embedding of science, its contextualization (*Nowotny et al.* 2001*; Nowotny* 2005; *Schulze-Fielitz 2005*) may be identified as a common ground, from a legal perspective these phenomena are better to be treated distinctly. They represent partially a reaction of the public to an ever closer coupling of science with other societal subsystems like politics, economy, law or mass media. No doubt, the governance structure of science is changing considerably and one expression is the "juridification" of the context (*Schulze-Fielitz* 2005). Law becomes medium and factor of the process of contextualization. If this diagnose is valid, the discourse under the headline "science and the public" is simplifying to a certain extent the phenomena which can be observed. Democratizing is not or at least not only about inclusion of the public in the scientific system or the observation of science by mass media, but it is about inclusion in very different societal subsystems. This results in complex institutional arrangements, coupling science, politics, law, economy and media. These complex institutional arrangements are seldom at the heart of the discourse of social studies of science. It tends to focus on particular elements of these arrangements, focus groups, consensus conferences and the like, or the general trends to be observed, thus underestimating the complexity, consequences and adverse effects of coupling different societal systems.

Coming to the subject of my paper, it is obvious that law has some experience regarding the inclusion of science and the public in institutional arrangements, especially in decision-making procedures. I will focus in this paper on scientific advice, and will in particular deal with risk decisions. Risk decision-making represents a good field of reference for the above mentioned institutional arrangements as we are confronted by a combination of science, politics, law, economy and the public. It enables us to mark some of the adverse effects to be observed when coupling different societal subsystems and the consequences which could be drawn from this. Scientific knowledge is by definition characterized by uncertainty, which causes specific problems for the legal system as well as science in being part of the decision-making process and including the public also is all but an easy task, as will be shown.

I will begin with some remarks concerning science and the public (II.), then show what risk decisions are about – particularly dealing with the repercussions scientific uncertainty has for legal approaches (III.) – and then I will address inclusion of the public (IV.) Arguing in favor of more differentiation and a subsequent recombination in the procedures I will particularly use the example of the deliberate release directive of EU as a background to describe the way in which science and public are arranged within legal procedures and what might be drawn from it (V.).

II. Science and the Public: Democratizing to Overcome Deficits?

If we identify the start of the debate on science and the public with the Bodmer report of the Royal Society in 1985 (*Bodmer* 1985), and leaving aside earlier American experiences dating back to the Sputnik shock or the long history of popularization of science (*Gregory and Miller* 1998; *Daum* 2002; *Stichweh* 2003), then the debate on science and the public has been ongoing for more than two decades. But after so many years of PUS initiatives still a crisis is diagnosed for example in the 3rd report of the Select Committee on Science and Technology of the House of Lords (House of Lords 2000). One of the reasons might be that public understanding of science was insufficient from the very beginning, this frequently owing to the fact that attempts to educate the public have been unidirectional and patronizing in nature and with the ensuing result that the public fails to understand science properly. Consequently the means are oriented to an enhancement of public scientific literacy: Enlightenment of the Public by scientists going public and painting a rosy picture of what could be done in the future with all the new scientific knowledge, and politicians eating a burger in public to assure that beef is safe to consume. Has the public – due to its experience with BSE, GMOs and the like – perhaps come to realize that those scientists going public are spin doctors for the scientific system?

Conceiving the public as illiterate and often as a pliant victim of distorting sensationalist media – both to be treated by proper scientific enlightenment – will lead to immune reactions of the public as well as the media system itself. It is not only that one can sometimes observe a gross misunderstanding of the public and its concerns and the way the media system works, but the public understanding initiatives often construct an image of the public and its concerns which prolongs the culturally dominant discourse of the cognitive authority of science (*Bucchi* 1998). And it misconceives the issue at hand. Various surveys in different European countries over the last decades have shown again and again that there is no such thing as a general mistrust of science (*Schulze-Fielitz* 2005; *Stehr* 2003: 314; House of Lords: Appendix 6). To the contrary, the public subscribes to the idea that science in general contributes to the common weal and that it is reasonable to put further hope in scientific development. What the public is suspicious about is more the handling of scientific knowledge in certain fields, especially when it comes to political and legal decisions being "legitimized" by science (e.g. risk decisions) and decisions about science conflicting with other societal interests and values (e.g. biotechnology). The 3rd Report of the House of Lords frames the problem – although too narrowly (which means too scientifically) – as follows: "Much of the public policy debate is confused by an assumption that the issues reverberating around science in the public domain, especially a variety of risk issues, can be reduced to a set of questions capable of objective and incontrovertible answer by scientific research. Most often, in truth, the issues are complex. Scientific understanding can contribute to a resolution of the issues, but only in partnership with judgements based on people's attitudes, values and ethics" (House of Lords, 2000: no 2.47).

A recently conducted comparative study (*Marris et al.* 2001) – backed by many others (*Shaw* 2002) – on behalf of the European Commission about public perceptions of agricultural biotechnologies in Europe revealed not only a lot of misleading conceptions that stakeholders, administrators and scientists have about the concerns. The public realizes quite well that due to the dominant science-based risk assessment discourse it is left with the false alternative of acceding to what scientists advocate as "real truth" or else be labelled as irrational and emotive, raising inappropriate and obstructive ethical objections against what is scientifically judged to be sound. The public seems not so much concerned about risk nor does the public advocate a zero-risk approach – often assumed by stakeholders. To the contrary, the public takes the existence of uncertainty for granted and assumes that there will always be unanticipated consequences and that this is not the fault of science. Given the inevitable ignorance, the purposes of and need for innovations are central concerns and it is obvious that they cannot be suitably addressed within a frame of science-dominated risk assessment. Due to uncertainty social, political and ethical concerns may become more important. Questions like: What is the sense of a certain innovation? What kind of knowledge is necessary, which questions have been considered? Have the relevant alternatives been taken into account? Is the institutional arrangement constructed in a such way as to provide appropriate measures for acquiring new experience and countering unanticipated consequences? Is there an adequate institutional frame to control activities and do those in charge have the necessary power to override those who would try to hide certain detrimental effects as long as possible, as seen in the BSE crisis? Are there enough independent scientists in an age where the relationship between university research and industry is becoming ever closer? One of the main problems seems to be a certain reductionism and reluctance of both the science system and the policy maker to accept the fact that in a differentiated society there will always be a plurality of perspectives and that the scientific view is just one of many.

Therefore it comes as no surprise that the aforementioned 3rd Report (House of Lords 2000) advocates a dialogue between science and society, or, to use the words of Helga Nowotny et al., society speaks back to science (*Nowotny et al.* 2001: 50 ff.; cf. *Miller* 2000). This represents a considerable change of perspective. In the science system, for the longest time no inclusive roles for the public have emerged. The audience for the scientists is composed of the scientists themselves – at least those who are part of the specific epistemic community. Historically this may be part of the contract with society for evolving as an autonomous system (*Luhmann* 1998: 624 ff.; *Stichweh* 2003; *Nowotny* 2005). Seen against this background, one may argue that many of the tendencies to be observed are an attempt to foster the integration of science into society by means of inclusion. Democratization, the new catchword for this process, aims at participation of the public also *within* the science system. The integration of lay knowledge and local expertise may be seen in this context. AIDS patients or other lay experts participating in research and funding procedures point in the same direction (*Kent* 2003). Advocating socially robust knowledge, which is produced in a plurality of different but formerly more or less separated arenas, as Helga Nowotny et al. do (*Nowotny et al.* 2001: 166 ff.), is the recipe for a better integration of science into society. No doubt, this attempt will lead to a kind of democratization at least in the sense that the definition and control of what constitutes reliable knowledge lies not only within the science system. Claims for more accountability, plurality of stakeholders in decision-making contexts, mentioned by Dominique Pestre in his paper (*Pestre* 2005), transparency, and participation point in a similar direction. The discussion about a scientific citizenship concentrates on many of the aforementioned aspects, trying to reconcile them with models of democracy (*Elam and Bertilson* 2003).

As mentioned above (supra I.), from the perspective of a legal scholar it seems important to note, that this is not only about a change in the relationship between science and the public, but it is also about inclusion of the public in the political and legal system, as far as science is subject of a political process or legal procedure or science provides knowledge for the decisions to be taken (*Trute* 2005). To a certain extent a closer coupling of science, politics, law and the public can be diagnosed (*Weingart* 2001: 115 ff.). This might be seen as part of the process of contextualization of science (*Nowotny et al.* 2001: 96 ff.) or as democratizing science in a broad sense.

If we take a look at the European level then we find a considerable – although ambivalent – change of perspective in the White Paper on Governance (Commission of the EC 2001), which advocates also greater inclusion of civil society. The Report of the Working Group "Democratizing Scientific Expertise and Establishing Scientific Reference Systems" (2001) can be seen as a reflection of this broader discourse, trying to develop rules for the generation, use and communication of expertise, and trying to reconcile scientific expertise and public participation in a new way and attempting to design a kind of "due process" or "due access": not only "democratizing" scientific expertise but also "expertizing" democracy. This claim for mutual adjustment poses the question of how to design it institutionally.

Following my argument that contextualization often results in complex institutional arrangements (cf. *Roqueplo* 1995 for a strong plea for institutionalization) coupling different societal systems I will, first of all, examine the interface between law and science and discuss some of the adverse effects and possible consequences.

III. Law and Science in a Regulatory Context: A Misalliance?

The law institutionalizes the frames not only for the inclusion of science and the public but sets rules for the generation of knowledge in a broader sense. This can be regarded as part of a legal construction of the (scientific) knowledge base of decisions. The inclusionary and exclusionary effects of the frame as well as the rules for the generation of knowledge influence what counts as knowledge/facts, at least within the legal realm.

I will focus first on the interface between science and law. Science is often seen as a cornerstone of risk decision, as a legitimizing factor and a means of avoiding arbitrary decisions – as the European Judiciary argues time and again. But what happens, when science provides only uncertainty?

1. What Is Special About Risk Decision

I will start by briefly sketching what risk decisions are all about and why they cause problems for the law as well for science when included in the regulatory context (cf. *Trute* 2003: 79 ff.).

a) First-Order Risks and the Problem of De-legitimizing the Law

In the legal sphere, the term "risk" covers different degrees of (un)certainty. According to decision theory and economic analysis, risk generally describes situations in which all possible outcomes are known in advance and where their relative likelihood can be expressed as probabilities: risk as a product of the probability of an event and the magnitude of possible losses. This is the classical and well known dimension of hazards or dangers, as it is called in the German legal tradition. At least since the Industrial Revolution the law had to deal with possible hazards (railways, steam engines, explosion of gas boilers etc.). The decision in cases like this is – *cum grano salis* – built upon experience and enriched on a case by case basis through the expertise of technicians such as engineers and through the building up of conventional guidelines as to which margin of probability was acceptable – this clearly with a bias in favor of technology. These conventional guidelines emerged by experience and are of course not merely an expression of pure scientific or technological knowledge but contaminated by normative assumptions about what counts reasonably as an acceptable margin of danger.

With the term "risk decisions" I am referring to situations in which the formal probabilities cannot be assigned to the outcomes (cf. *Harremoes et al.* 2002: 188) due to a lack of empirical data or a theoretical basis (uncertainty) or a situation where we have to deal with unknown possibilities (ignorance). This applies in circumstances where no basis for assigning probabilities to specific outcomes exists and where a complete set of outcomes is also problematic. This may root in the novelty of the activities concerned, the complexity or variety of the factors and their context (atomic energy, GMOs, climate change).

In situations like this the legal system faces greater problems as the role of science in legitimizing at least the factual basis of a decision seems often to fall short. Scientific knowledge is temporary, uncertain, contested, and mixed with assumptions and values. For different reasons, constitutional as well as political (*Trute* 2003), the legal system cannot ignore uncertainties below the traditional level of danger and now faces the problem that science cannot provide the necessary knowledge to build a legal decision upon, at least not in the way it was done before. There are no stable conventional guidelines legal decisions can be built upon and consequently decisions like this are in most cases highly contested. Is not legal *un*certainty the result when a decision is built on the quicksand of provisional scientific knowledge?

Under the umbrella of the precautionary principle (*Harremoes et al.* 2002), which is the principle normative frame for handling uncertainty, the risk law (*Trute* 2003) often refers to notions like "state of science and technology" in granting a license for a technology or pharmaceutical product or the like so as to assure that the product or technology is safe enough. This reference would seem to shift the responsibility for normative decisions to politically unaccountable science, which then pronounces on what it does or does not consider to be state of the art (*Bora* 1996: 392). As the state of science itself is uncertain and probably value-laden and, as science cannot answer the normative question of what constitutes acceptable risk, both science and law suffer. Science can be criticized for being political and thus overreaching its prerogatives, and legal decisions might be de-legitimized as being scientifically underdetermined and allowing for this shift in responsibility (*Weingart* 1999: 158). Either the uncertainties are hidden by the claims of science to render answers to those questions which are obviously beyond the scope of scientific knowledge or uncertainty can be revealed at the expense of the scientific authority's legitimacy. Science would then become politicized, and politics and law would become "scientificated". Thus adverse effects are to be expected in any case. The public would soon come to realize that scientists act as interested parties in a decision-making context, either on behalf of the administration or of private actors. This undercuts their status as cognitive authorities and the law then faces the problem of becoming de-legitimized. The function of the law to provide a certainty of expectation, which is one of its core functions, seems to be diminished, as sociologists of law consistently argue. The trust of the public in two important institutions of modern society seems thus to be diminished.

b) Second-order Risks and Balancing Approach

If law attempts to regulate these first-order risks (scientific risks related to the subject to be regulated) it generates a new kind of risk, that of regulating a risk which may have unenvisioned detrimental effects. This is linked to complexity. We cannot know the future path of science, what options will be chosen by economic actors, which innovations will be implemented and how they will affect our present calculation of risk in the future, how politics will react and what the courts will decide. Every decision as well as every *non*-decision has some kind of risk attached. Thus there is not only scientific uncertainty about the subject to be regulated but also the risk related to regulation as such. This might be called second-order risk (*Scherzberg* 2004: 220 f.; *Trute* 2003: 78 ff.). To deal with it adequately the law must become reflexive and anticipate the consequences of regulation and know that it might *un*knowingly regulate false positives as well as false negatives, or to put it differently, it regulates activities which might in the future be seen as harmless but with detrimental effects on useful innovations, or it might or might not adequately regulate activities with unanticipated detrimental effects to other goods in the future. Regulation of this kind is necessarily contextual and results in a balancing approach. Case by case the available knowledge has to be assessed in the light of possible consequences of a decision or non-decision and the legal values at stake.

c) The Burden of Politics

The consequence is a shift of political responsibility to the administration to allow for this balancing approach in each case, the results cannot sufficiently be prescribed by a general law. The administration has to be given flexibility and broad discretion in administrative decision-making, at least to a certain extent and of course dependent on the subject to be regulated. And as the decision is scientifically underdetermined, political, ethical and moral aspects become more important and have to be taken into consideration within the administrative procedure. But this goes at the expense of legal certainty.

2. Renormativation, Proceduralization, and Participation

Administrative decision-making is obviously not in a comfortable situation: delegitimization due to scientific uncertainty, politicization due to the context-dependency and a possible loss of legal certainty. There are at least three interrelated tendencies to cope with these adverse effects of a close coupling of science and administrative decision-making within a normative frame: renormativation, proceduralization, and participation. This results in an increasing complexity of the normative frame.

a) Renormativation by the Courts

Although "de-legitimization" may have been an appropriate term during the early stages of discussion regarding risk within the German legal context – say, up until the end of the 1970s – with the so-called Wyhl-case (Bundesverwaltungsgericht 1985: 312 ff.) dealing with a nuclear power plant that was planned to be built near the famous vineyard region of the Kaiserstuhl, things began to shift in a manner that I call "renormativation," which might be regarded as a re-entry of normative criteria in the decision-making context. The underlying rationale was to develop normative criteria and institutional arrangements that allowed for dealing with uncertainty and the frame was the precautionary principle, which could be seen as a kind of normative interface of science and law. Summarizing the findings, we can observe that uncertainty of science *as such* became the input-data for the legal decision. The interesting question then from the standpoint of the legal system is *which normative* conclusions should be drawn from *this* fact. Probably different from the American regulatory context, the court does not – or only within narrow limits not to be discussed here – deal with details of the scientific discourse, nor does it allow for a shift of administrative responsibility and accountability on the shoulders of science, but takes the fact of uncertainty as a starting point for the normative assessment of what might be normatively justified. Differentiation of risk-management as the normative side and risk assessment as the scientific side of things thus become complementary aspects of risk decision-making.

The above mentioned reflexivity also points in the direction of a renormativation, as the balancing is necessarily a normative aspect of the decision making. This, as well as other aspects, might be read as a *partial separation* of normative criteria from the scientific expertise which tries to avoid the use of science in an instrumental way and – at least to a certain extent – tries to shield science from political interference. And it assures political as well as judicial accountability. This seems to point to a demarcation model (*Liberatore and Funtowicz* 2000: 148 f.) between science and law, leaving unaddressed the possibility of framing, hidden assumptions and values in scientific knowledge. As we will see later on, the emphasis on a partial separation is compatible with procedures that address these crucial aspects.

b) Proceduralization

All this goes hand in hand with the second aspect, proceduralization. Due to first- and second-order risks, the law must provide for a scheme, which allows for flexibility, and temporality as well as the enshrinement of learning capacities within a legal frame in the sense that legal decisions are not seen as the final point of a decision-making process but rather allow for a questioning of the assumptions the decision was based upon in the light of new knowledge, changed circumstances and the like (*Ladeur* 2003). This approach includes the dimension of time into the legal frame (*Trute* 2003: 98 ff.) and allows for permanent generation of new knowledge. Proceduralization is thus to be read as a normative scheme that provides for flexibility on the one hand, and legal certainty on the other. Instead of determining the administrative decision on substantive grounds an elaborated procedure for finding the decision is designed as a compensation for the weak substantial determination, enhancing the legitimacy of the decisions to be taken.

c) Participation

If the decision is contextualized in the aforementioned manner, if there is a shift of responsibility to administrative procedure, then inclusion of the public becomes important as a kind of compensation for the lack of parliamentary guidance and control, thus helping to enhance legitimacy and subsequent acceptance as well as improving the knowledge base. This may thus come in the guise of democratization, as a complement for granting a broad discretion to the administration or as a tool for enhancing information capacities and the acceptance of decision-making in view of the uncertainty.

IV. Including the Public: Experiences and Consequences

Now the stage is set for a discussion of the second interface, the inclusion of the public in risk decision-making. Inclusion seems to be a value in itself, at least from the view of democratic theory. It can take various forms due to the social system in question and the subject at stake. Experiences with the inclusion of the public in legally determined procedures reveal at best a mixed outcome; enhancing inclusion – as will be seen – does not necessarily foster integration and acceptance of a decision to be taken.

1. Different Forms of Inclusion

The legal system distinguishes between participation due to competing legal principles – for example the right to privacy of those being affected – and confidentiality and the like. Participation therefore takes place in very different forms and along very different spectrums. It may range from those being addressed to those being affected by a decision to the public at large, raising objections can be made in written form or at an oral hearing. Scope and form of inclusion influences the knowledge base of the decision to be taken.

At a minimum, participation of the public should inform both the decision-maker and the public, should monitor both procedure and decision-making, protect the rights of those being affected, represent interests, and make for legitimation and subsequent acceptance of the decision. If we consider the contextual character of risk decision-making and the intrinsic problems of scientific knowledge (framing, hidden assumptions, value-laden character) and if we subscribe to the idea of democracy, then broad inclusiveness would be appropriate, a kind of deliberative arena to discuss the relevant aspects, be they normative or scientific. This would require an oral hearing so as to allow for reasoned argumentation. And indeed, environmental as well as urban-planning law provides for this form, albeit with mixed outcomes.

2. Experiences with Oral Hearings

This form of inclusion might work if there were to be no principal objection on the part of the public or NGOs to the activities in question. But this is normally not the case, particularly in the instance of technologies and hotly contested issues such as GMOs. Pertaining to the debate surrounding GMOs, and within the older German legal frame, the outcome might be summarized as a politicization of the law and a juridification of politics – just two sides of the same coin. Politicization of the law means the introduction of moral, ethical and general political questions within the given legal framework in order to transcend it. Questions like: Do we need GMOs? Is it natural? Is it reasonable to foster such a technology in light of the unknown risk? Once a political decision is taken to allow – at least in principle – for the activities and it is then transformed into law, such probing questions transcend the legal frame of the procedure. These are immediately parried by the administration with the argument that the administrative procedure is not designed to compensate for presumed deficiencies in the political process and that the administration can only act within a given framework without transgressing the boundary between legal and illegal decisions. This is perceived as a powerful juridification of politics by those who fundamentally oppose such technologies. A closer look at the arguments and strategies being used also reveals something that might best be characterized as a clash of different discourses, each conceiving the procedure, its aim and its function in a different way. Experts (including counter-experts and lay people) have expectations divergent from those using the procedure as part of the political process, sensationalizing in hopes of public attention via the media system. Those who perceive the procedure as a part of the ongoing political struggle against GMOs soon come to realize that the experts are either part of the political power or share the interests of those who object to the new technologies. The result is not resolution of questions in a deliberative discourse but frustration of all those participating in the procedure; the outcome is not legitimacy and acceptance of the decision but ongoing protest. Inclusion through participation does not necessarily enhance integration; in fact, the contrary may be true (*Epp* 1998).

3. Some Reasons Why

There might be different theoretical explanations why this happens, either you argue along the lines of systems theory or you follow a discursive model. For those familiar with systems theory, all this should come as no surprise. Embedded in a complex theory of systems *Luhmann* distinguishes between risk and danger (*Luhmann* 1991: 37 ff.). Both refer to a future damage, but risk is attributed to a decision and danger is something that happens to someone.

This leads to the distinction between those who decide and those who are affected by a decision. The first calculate the risk of a decision, the latter observe themselves as potential victims of a decision. Both use incongruent and often hardly reconcilable perspectives, as frequently observed in disputes on environmental matters (*Bora* 1996, 1999; *Epp* 1998). Risk theory for a long time has given us not only enough empirical evidence for the relevance of this distinction, but also that risks are treated differently whether they will be attributed to own decisions and behavior or to that of another party (*Japp* 2000: 65 ff.). If this is true, we will have to be skeptical as far as the participatory approach to risk decision-making to solve conflicts with those who are affected by risk decisions is concerned. In a more pragmatic sense there will be negotiated solutions possible, but it could be an illusion to subscribe to the idea that consensus might be the ultimate solution to highly contested risk decisions. In the end, the legal discourse necessarily takes precedence. It follows the artificial reason of law, which allows normally not for political and ethical reasoning in administrative procedures, at least as long as the legal standards are not opened to such arguments. Although empirical evidence supports the theoretical conception (*Bora* 1999; *Epp* 1998), it perhaps misconceives the function of the administration as also being part of the political system (*Trute* 2004: 302 ff.) and it underestimates those ambiguities enshrined in the law (for a more general discussion *Grant* 2004) so as to allow for reinterpretation likewise in the context of legally determined procedures. However, legally determined procedures of course place severe constraints on deliberation, notwithstanding the problem of bridging the gap between those who decide and those being affected by a decision.

I will not go over this ground in detail (*Trute* 2003), instead referring to the results of a recent study in the *Oxford Journal of Legal Studies* (*Black* 2001). Julia Black tried to develop what she calls a thick conception of proceduralized regulation (meaning a conception of deliberative arena in administrative decision-making). Although very sympathetic to this concept, she concludes that the obstacles can be extremely daunting within a regulatory context. She points out that there are differences in the discourses that are hardly compatible within the regulatory context minus further institutional devices such as "translators" – meaning actors able to bridge the gap between discourses and make transparent the different cognitive frames of regulators, scientists and other participants. To balance the various skills and resources, mediation needs to be provided so as to be aware of the inclusionary and exclusionary effects of problem-definition and agenda-setting and the like. And a strategy of dispute-resolution needs to be available so as to reconcile conflicts of interest. Although not easily compatible with the framework of administrative decision-making and those devices assumed to be at hand, other constraints are difficult to overcome: These include the possibility of strategic interaction, the need to take a final decision in terms of legality (which is also the need to consider the legally protected interests of those seeking a decision), and the need to take a final decision in time (which is not only a matter of efficiency but of legal guarantees and constitutional values, last not least that of democracy). The skeptical conclusion is that it may well be that a thick conception of proceduralization must abandon the hope of normative consensus; that deliberative democracy seeks and aims instead for a discursively produced coordination rather than integration (*Black* 2001: 58). Different discourses, be they scientific, political, legal or ethical are not easy to reconcile within one deliberative arena. Even when advanced institutional devices to bridge the gap between them are at hand, the argumentation follows the lines of the artificial logic of the respective discourse, thus establishing considerable high barriers for integration. Therefore it is probably no accident that the above-cited EU Report on "Democratizing Scientific Expertise" (2001: 7) – although sympathetic to enhanced participation and to inclusion of different types of knowledge – is well aware of possible trade-offs with regard to credibility, legitimacy and efficiency of expertise and the constraints of the policy process and administrative procedure. Democratizing expertise is therefore understood as guaranteeing "due process" in the way expertise is developed, used and communicated.

4. Some conclusions

Obviously democratization does not necessarily go along with consensus and integration. To the contrary it might produce protest and mistrust. Again an increased complexity of the frame might ease the burden without solving it principally. My suggestion is to split the procedure and recombine it in a different way. Older models of coupling science and participation in one legally determined procedure mingled too much within one procedure, which results in an unstructured complexity. To split the procedure allows for a recombination, which assures an adequate complexity as well as high level of flexibility. I draw my claim for differentiation and recombination from experience in other fields, such like telecommunications and environmental law. To deal with uncertainty, there – first of all – has to be an adequate complexity of the legal frame, so as to allow for learning in the light of uncertainty, and – second – there should be separation of risk-management and risk- assessment procedure. This would allow for participation on the scientific as well as the normative side. To overcome the obstacles of the concrete decision-making procedure, it might be useful to create a more strategic or intermediate level between the general law and the concrete administrative procedure. This would allow for broader forms of participation on the strategic level and a more narrow form at the level of decision making. Of course, this model will not overcome the burden of politics to assure the legitimacy of the principle acceptance of a technology.

V. Differentiation and Recombination: The Example of the Deliberate Release Directive (18/2001/EC)

To give you an example of how it works or might work, I refer to the EU-directive on the deliberate release into the environment of genetically modified organisms (2001/18/EC). It has to be seen together with the regulation on establishing the European Food Safety Authority (Regulation (EC) No 178/2002). Both can be understood as an attempt to learn from the BSE crisis and to foster a broad institutional change.

1. The Burden of Politics: Defining the Level of Protection, Envisaging a Learning Procedure

Beginning at the policy level, it is up to the competent political authorities to define the chosen level of protection and to design a learning procedure to deal adequately with uncertainty.

a) The Level of Protection

The level of protection has to be chosen within legal confines, e. g. a high level of protection enshrined in the EC Treaty (Art. 174), which gives only a broad frame that leaves considerable discretion to the particular authority. In the EU directive it is defined as the obligation *to avoid adverse effects on human health and environment* by the deliberate release of GMOs. Note: It is not about risks, significant risks, negative impacts and the like, to be defined by the scientific discourse, but rather about what normatively counts as an adverse effect. Everything that may infringe this standard is normatively qualified as an adverse effect. This includes the decision that GMOs could be deliberately released, but only to the extent that no adverse effects are to be expected. This excludes further deliberations about the need for and purpose of GMOs, one of the public's main concerns. These concerns are left to be dealt with by the political process before a decision to adopt a precautionary frame is taken. It is important to note that some of the devices for deliberation, like TA, consensus conferences, focus groups are institutionally to be located at this policy level.[1] This backs our claim that "democratizing science" is a topic referring to

[1] Although one can safely state, that the impact on the political system has not exactly been overwhelming (*Abels and Bora* 2004). But there is obviously a need for new forms of inclusion at this level, which might ease the burden of the subsequent administrative process.

different societal systems and therefore covers very different institutional arrangements.

Coming back to the level of protection the notion "adverse effects" leaves broad powers of discretion to the administration with regard to what actually counts as "adverse". Therefore a procedure is needed that provides for the generation of knowledge and standards (supra II. 2 b).

b) Institutionalization of a Learning Procedure

The deliberate release directive is a fine example for proceduralization because it enshrines the precautionary principle within the procedure itself. This not only in the sense that measures based on that principle can be undertaken, but in the sense that the design of the framework in itself is an expression of the precautionary principle: the institutionalization of a learning procedure so as to provide for increasing knowledge in the application of the frame. It comprises, first of all, a *case-by-case* and a *step-by-step* approach. The case-by-case approach facilitates a mandatory environmental risk assessment (e.r.a.) of every single release, whereas the step-by-step approach provides for a line of development of GMOs by evaluating the impact of releases in decreasing steps of containment (from greenhouse experiments to field tests to complete market approval), and ideally there should be market approval only after the greater knowledge accrued through these gradual steps. Each case and each step is intended to provide new information, thus enhancing the relevant body of knowledge available. It is accompanied by compulsory monitoring, reporting and even possible revision. The e.r.a. is to be done by the applier, forcing disclosure of the knowledge base, which is then scrutinized by the competent authority to see if it is in accordance with standards defined in the directive. Compared to the frame used formerly in dealing with the aforementioned hazards of the classic type, this legal frame is not built on conventions generated by engineers and scientists in heuristic ways but rather aims at generating knowledge and conventions by means of a specific design of the procedure, which are subject to discussion of the assumptions they are built upon. Of course, the directive provides also for public consultation in this procedure, including, where appropriate, consultation with groups.

2. Institutionalization of a Knowledge Management Infrastructure

At the core of the "scientific side" lays the institutionalization of an infrastructure of knowledge generation, exchange and clearance, a kind of a knowledge management infrastructure. Two elements are important, the institutionalization of an independent scientific authority, also to be found in many other areas of risk decision-making (e.g. chemicals, drugs) and the creation of a knowledge exchange network comprising scientific bodies of the member states and the EU (for an analysis also of possible adverse effects of this regulatory regime see *Buonanno et al.* 2001).

The most advanced scientific advisory body is the ESFA, an independent authority designed to provide scientific advice in the field of food, including GMOs. It is the central monitoring board in the network. Although providing opinions for the decisions to be taken on the European level, it also aims at generating knowledge in this field on a more strategic level. Both the authority and the network represent the institutional part of the legal construction of what counts as relevant scientific knowledge/facts. Following the demands of the European Judiciary it must be based on the principles of independence, excellence and transparency.[2]

a) Independence

Independence comprises different dimensions, politics and societal interests. Creating an independent agency implies always the dilemma of possible trade-offs between political and judicial accountability on one side and independence on the other side, a crucial matter in this case. If a high degree of independence is granted, additional checks would be appropriate, which can be seen e.g. in the inclusion of the public, transparency and the exchange networks itself.

Without going into the details, the regulation grants a considerably high degree of independence vis-à-vis politics. The management board, appointed by the European Council, is composed in a way that also consumers and other interests in the food chain are represented (Article 25 (1) Regulation (EC) No 178/2002). It appoints the Executive Director as well as the scientists, on the basis of an open appointment procedure, eliminating direct political interference. Members are obliged to act in the public interest and a procedure for declaring conflict of interests is established, including publication of the results. Of course, this is not a strong guaranty, as potential conflicts are not excluded principally. But there will always be the dilemma, that those having the relevant knowledge have more or less intense relationships to interested parties. A trade-off between excellence and independence is not easy to balance. Plurality of the scientific panel, the exchange network and obligations for transparency might be seen as a check (*Schulze-Fielitz* 2005).

[2] European Court of First Instance (ECFI) Judgement of 11.9.2002 (Case T –70/99) para 172; ECFI Judgment of 11.9.2002 (Case T-13/99) para 172, 323.

b) Excellence

Harder to deal with is the claim for excellence. Of course, the law could easily demand for high excellence, but it is hard to assure it in practice (*Pestre*). To count on scientific reputation is one solution, not easy to assess from outside the scientific system. But excellence is not only to be counted in terms of personal reputation. The institutional dimension has also to be taken into account to assure that the results are excellent. Here the knowledge exchange network comes into play. The agency works as a centre of reference in a network, comprising other scientific bodies of the member states. First, a high level of diversity is enshrined in this frame. The exchange with the national scientific bodies – partly within a formal procedure, partly as a general task – assures that knowledge is used, that derives from different institutional and cultural settings, be this public or private, university or regulatory science, be it European or national. Although a weak point might be that the composition of the scientific panels, obviously a crucial point as it influences the results, is not laid down in the directive but left to the authority, there are other important achievements: a procedure for clarifying contentious scientific issues and the publication of the results including remaining open questions and the possibility to give minority votes, also to make public. It is fair to argue, that this institutional setting at least increases the probability of excellence, as all relevant aspects have a fair chance to be included.

c) Transparency and the inclusion of the public

A high level of transparency is guaranteed, including the publication of agendas and minutes of its board and panel meetings as well as making them open to the public. Consumer and other interest groups are represented on the management board. There is a publication of the scientific studies and opinions. A stakeholder forum is conceived, and the authority has the competence to hold public hearings on scientific issues. The public can thus be included in diverse ways on this scientific side, without suffering the constraints of the concrete administrative procedure. It is important to note that the EFSA adopted a policy for consultation, which guarantees a high level of consultation in principle matters of risk assessment. Nothing prevents the EFSA to carry out TAs, stakeholder conferences and the like, to provide for a broad inclusion of the public.

3. Recombination in the Decision-making Procedure

The directive in combination with the regulation on EFSA provides for a clear separation of normative risk-management and risk-assessment. But expert knowledge must obviously be transformed into the normative language of the directive and the respective national laws, which then transpose the directive. This is done in the decision-making procedure for the application for a deliberate release. As far as I can see, there are basically two aspects to note: the normative formation of the environmental risk assessment, and a kind of concept that couples scientific with normative language. This recombination is sometimes, especially in older frames, hidden to a certain degree and thus leads scholars to take a direct look at argumentation strategies, trading zones, negotiations and deliberations. Although there is – as we will see – such a trading zone, this perspective does not take into account that the scientific procedure as such is at least partially designed by the law. Again, there is a *substantial* and a *procedural* side.

Indeed, risk-assessment may be seen as a scientific procedure, employing the methods and practices of the scientific community but transformed in a regulatory framework. The procedure as such is attended by rules which set the general principles, methodologies, standards and steps to be met.[3] My assumption is that the generation of knowledge is affected by this frame, since it is not one that allows for the "normal scientific procedure" to take place. Rather, it prescribes certain particular steps and methods, the result being that the knowledge produced is different from that generated by a normal scientific discourse. This is part of the legal construction of scientific knowledge (The normative frame can thus be evaluated, whether it facilitates or inhibits the production of reliable knowledge; cf. *van Zwanenberg and Millstone* 2000).

Secondly, in the end the e.r.a. must substantially meet the normative standards whose underlying principle is *the adverse effect on health and environment*. But when can we speak of adverse effects? My hypothesis is that there are conventions at work which function as a kind of yardstick. One prominent example may be the "natural situation" itself. This convention is used by both sides as a reference or scheme that is not in itself fixed, as our knowledge and images of nature do necessarily change. In the case of the directive, we are dealing with *familiarity* and *substantial equivalence,* which function within a comparative analysis between GM crops and non-GM counterparts as a substantial measure of whether there are adverse effects or not. The concept of familiarity is based on the fact that most genetically modified organisms are developed from organisms which are well understood, and the assumption is that traditional knowledge can be used for an assessment of adverse effects as long as there is substantial equivalence between a GMO and a well-known non-GMO. Of course, this is using nature as a comparator and there is neces-

[3] See Directive 2001/18/EC Annex II, which sets the frame, by giving the general principles, methodology, steps and conclusions.

sarily a difference between GMO and non-GMO; but it is a distinction which allows the similarities and differences to be identified and evaluated.[4] This convention functions as a kind of interface between the scientific assessment and the normative procedure.[5] The law uses it as a substantive standard which has to be met and science uses it as a description of a scientific strategy of comparison to provide robust knowledge about the safety of a certain GMO.[6] This convention is obviously value-laden and can be questioned on scientific, ethical and normative grounds. And it is important to note that this convention is not explicitly generated in a transparent procedure but has evolved over time. As it is of more general meaning, it should therefore be open for deliberation on a more general level so as to make both it and its underlying assumptions explicit.

4. The Claim for a Strategic Concept

But, and this is one of my reservations, there is no such intermediary procedure between the general law and the administrative decision on a concrete level available, to make this conventions explicit and subject to a broader discourse. In a more informal way the EU uses guidelines etc., but a formal procedure dealing with this more general aspect to develop such a concept is not envisaged. Although there is enough evidence and experience in other fields such as telecommunications law and environmental law, to mention only two fields, that scientific models and assumptions, the hidden frames and knowledge gaps as well as general normative principles of regulation could be subject to a broader discourse in risk law, if they are distanced from the concrete case, it is seldom used in this field. And this intermediary level would allow for broader inclusion of the public, avoiding the constraints of the concrete decision making procedure. The consequence is that again all these principle aspects are to be discussed in the concrete administrative decision-making procedure with the aforementioned results. And this goes also at the expense of legal certainty as the danger of arbitrariness of the risk decision is not banned.

---

4 See Guidance Document of the Eurpean Commission for the risk assessment of Genetically Modified Plants and Derived Food and Feed, 6-7 March 2003 p. 7 f.; the opinions of the competent scientific panel of the EFSA are using this standard – as far as I can see – in every case.

5 These examples are drawn from *R. v. Schomberg* (2004), who calls them transformable normative Standards. Although agreeing in principle I want to stress their function as interfaces which allow the scientific system and the legal system to speak – so to say – the same language.

6 FAO/WHO Report of a Joint FAO/WHO Expert Consultation on Foods Derive from Biotechnology, Safety aspects of genetically modified foods of plant origin, 2000, p. 20.

VI. Conclusions

But despite this critic it is fair to note there has been considerable change with respect to expertise in the aftermath of the BSE crisis and the protest against GMOs, at least at the EU level. The creation of the knowledge management infrastructure and the institutionalisation of a learning procedure are important steps, as well as the inclusion of ethical aspects I left aside in my paper. This is at least a step to more accountability by differentiating procedures, institutionalizing different arenas for the inclusion of the public. The result of this learning process of society is a complex governance structure. Whether or not this would lead to more acceptance and integration remains to be seen and adverse effects as well as unanticipated consequences are to be expected. The law can only provide for procedures which guarantee the inclusion of all affected interests, the transparency of the process, and – let us hope – a just decision. It does not promise acceptance or integration, only legality.

The banner of "democratizing science" covers very different aspects, the inclusion of the public (lay persons) in the research process, inclusion of the public in political and administrative procedures dealing with scientific subjects or the use of scientific knowledge, the integration of science and society or the observation of science by the media, to mention just a few of them. Broad and ambiguous in its meaning the term tends to simplify the underlying problems and to create false distinctions and dichotomies. Enhancing participation, positive as it may be seen from democratic theory, is not to be seen as a simple device for integration of science and society. There is enough experience for disillusion. Modern societies are highly differentiated and reintegration is both a complex and enduring task. From the perspective of a legal scholar "democratizing science" in the broad sense refers to complex institutional arrangements coupling different societal subsystems, each operating along the lines of its respective rationality. To take difference as a starting point may be a good counterweight against overoptimistic expectations of fostering integration of science and society by enhancing public participation. The architecture of contextualization is necessarily complex and allows not for easy solutions.

Bibliography

Abels G, Bora A (2004) Demokratische Technikbewertung. Bielefeld

Black J (2001) Proceduralizing Regulation: Part II, Oxford Journal of Legal Studies 21: 33–58

Bodmer W (1985) Public Understanding of Science. Royal Society, London

Bora A (1996) Inklusion und Differenzierung. Bedingungen und Folgen der „Öffentlichkeitsbeteiligung“ im Recht. In: van den Daele W, Neidhardt F (eds) Kommunikation und Entscheidung. Berlin, pp 371–406

Bora A (1999) Differenzierung und Inklusion. Baden-Baden

Bucchi M (1998) Science and the Media. Alternative routes in scientific communication. London

Bundesverwaltungsgericht, Entscheidung vom 19. Dezember 1985, Amtliche Entscheidungssammlung 72: 300–332

Buonanno L, Zablotney S, Keefer R (2001) Politics versus Science in the Making of a New Regulatory Regime for Food in Europe, European Integration online Papers, vol 5, no 12

Commission of the EC, European Governance, A White Paper, COM (2001) 428 endg., Brussels 2001

Daum A (2002) Wissenschaftspopularisierung im 19. Jahrhundert. München

Elam M, Bertilson M (2003) Consuming, Engaging and Confronting Science. The Emerging Dimensions of Scientific Citizenship, European Journal of Social Theory 6 (2): 233–251

Epp A (1998) Divergierende Konzepte von „Verfahrensgerechtigkeit“. Eine Kritik der Procedural Justice Forschung. In: Wissenschaftszentrum Berlin, Discussion Papers FS II, pp 98–302. Berlin

Grant CB (2004) Uncertain Communications: Uncertain Social Systems. Soziale Systeme 10: 217–232

Gregory J, Miller St (1998) Science in Public. Communication, Culture, and Credibility Cambridge

Harremoes P, Gee D, MacGarvin M, Stirling A, Keys J, Wynne B, Guedes Vaz B (eds) (2002) Precautionary Principle in the 20th Century. Late Lessons from early Warnings. London

House of Lords (2002) Select Committee on Science and Technology, 3rd Report. Science and Society, London

Japp KP (2000) Risiko. Bielefeld

Kent J (2003) Lay Experts and the politics of breast implants. Public Understanding of Science 12: 403–421

Ladeur KH (2003) The Introduction of the Precautionary Principle into EU Law: A Pyrrhic Victory for Environmental and Public Health Law? Decision-Making under Conditions of Complexity in Multi-Level Political Systems. Common Market Law Review 40: 1455–1479

Liberatore A, Funtowicz S (2000) ‘Democratising’ expertise, ‘expertising’ democracy: what does this mean, and why bother? Science and Public Policy 30 (3): 146–150

Luhmann N (1990) Die Wissenschaft der Gesellschaft. Frankfurt

Luhmann N (1991) Soziologie des Risikos. Berlin

Marris C, Wynne B, Simmons P, Welden S (2001) Public Perceptions of Agricultural Biotechnologies in Europe. Final Report of the PABE Research Project, Brussels

Miller St (2000) Public Understanding of Science at the Crossroads, paper prepared for conference on: Science Communication, Education and the History of Science. Royal Society, London

Nowotny H, Scott P, Gibbons M (2001) Re-Thinking Science. Cambridge

Nowotny H (2005) The Changing Nature of Public Science. In this volume: 1

Pestre D (2005) The technosciences between markets, social worries and the political: how to imagine a better future? In this volume: 29

Report of the Working Group (2001) "Democratizing Expertise and Establishing Scientific Reference Systems". Brussels

Roqueplo P (1995) Scientific expertise among political powers, administrations and public opinion. Science and Public Policy 22: 175–182

Scherzberg A (2004) Risikosteuerung durch Verwaltungsrecht: Ermöglichung oder Begrenzung von Innovationen. Veröffentlichungen der Vereinigung der Deutschen Staatsrechtslehrer 63: 214–258

Schulze-Fielitz H (2005) Responses of the Legal Order to the Loss of Trust in Science. In this volume: 63

Shaw A (2002) "It just goes against the grain." Public understandings of genetically modified (GM) food in the UK. Public Understanding of Science 11: 273–291

Stehr N (2003) Wissenspolitik. Frankfurt

Stichweh R (2003) The multiple public of Science: Inclusion and Popularization. Soziale Systeme 9: 210–220

Trute HH (2003) From Past to Future Risk – From Private to Public Law. In: European Review of Public Law 15: 73–103

Trute HH (2004) Methodik der Herstellung und Darstellung verwaltungsrechtlicher Entscheidungen, in: Schmidt-Aßmann E, Hoffmann-Riem W (eds) Methoden der Verwaltungsrechtswissenschaft. Baden-Baden, pp 293–327

Trute HH (2005) Comment: from a Legal Perspective, In this volume: 53

Weingart P (1999) Scientific Expertise and political accountability: paradoxes of science in politics. Science and Public Policy 26: 151–161

Weingart P (2001) Die Stunde der Wahrheit. Weilerswist

van Zwanenberg P, Millstone E (2000) Beyond Sceptical Relativism: Evaluating the Social Constructions of Expert Risk Assessments, in: Science, Technology & Human Values 25: 259–282

European Law

Directive 2001/18/EC of the European Parliament and of the Council of 12 March 2001 on the deliberate release into the environment of genetically modified organism, OJ (EC) No L 106, 17. 4. 2001, p 1.

Regulation (EC) No 178/2002 of the European Parliament and the Council of 28 January 2002 laying down the general principles and requirements of food law, establishing the European Food Safety Authority and laying down procedures in matters of food safety, OJ (EC) No L 31, 1.2.2000, p 1.

Free Access to Research Findings and Its Limitations

Eberhard Schmidt-Aßmann

I. Free Dissemination of Information: An Action Principle of Science and a Subject of the Law

Science has a distinctly public character: research results are intended to be published, and scientific knowledge is meant to be made accessible to the public on a permanent basis. Publication is required to validate research findings, and transparency is needed to create trust among members of society. Public access to information is a means of securing societal wealth, both now and in the future. In this sense, a free dissemination of information constitutes a vitally important action principle of science (*Weingart* 2001: 221 and 230).

Yet is this not a model from ages past (assuming there ever was a self-sufficient, "pure" science)? Did not science long ago adopt the modes of behavior common to industry and politics and gamble away its ethos of publicly accessible information (*Nowotny et al.* 2001)? Objections to the public-access paradigm come from a number of corners, especially military or pharmaceutical research. The ambivalent role of intellectual property rights also must be taken into account. Case studies of publications that have been deliberately suppressed or prohibited by commissioning parties make for an even bleaker picture. And new developments in patent law reflect massive efforts to transfer knowledge (even fundamental knowledge) from the public to the private domain (*Nowotny*, in this vol.).

Legal examinations should not confine themselves to merely describing such pathologies. If constitutions (including Germany's *Grundgesetz*, the EU's Charter of Fundamental Rights, as well as a number of recently drafted constitutions in other European countries) guarantee academic freedom (*Groß* 1992), societies must study the extent to which the instruments of law, among others, can protect and, if necessary, reinstate this freedom. For the underlying *normative approach*, sociological de-differentiation theses, as descriptive as these might be, must not be regarded as the conclusion but rather the starting point in the quest for possible counter measures. A broad or even absolute appropriation of science by industry or politics is in nobody's interest, since the sub-systems of a society are only functional when they are able to provide their specific services. And for this, autonomy is required.

It goes without saying that jurisprudence regards this independence as a *relative autonomy* distinguished by a variety of interrelations with other social subsystems. The primary task of law is to structure these interrelations. This also applies to the subject at hand. In this paper, we will not only examine the legal regulations directly or indirectly governing publication and public access to research findings, but also describe the anticipated effects of these regulations as steering instruments.

The free dissemination of information as an action principle of science is not identical with the contents of legal regulations relating to public access to information. The law must balance out a number of competing interests. For instance, the dictate of public access may at times stand in opposition to the idea of protecting privacy or corporate secrets. This requires that the different interests be analyzed, systematized and given fair weighting. Such a systematization can be particularly conducive to pointing out loopholes in legal protection. Any portrayal of the legal situation must consequently incorporate political considerations.

Of course, laws can only make a *partial contribution* to helping science stick to its action principles and safeguard its relative autonomy. Even today, publication and free access to information continue to be the norm in most disciplines of science – which cannot go unrecognized despite all critical observations. The mechanisms of scientific *self-regulation* are just as important as legal regulations. By giving greater attention to this topic in the future, the law can lend its support to those forces immanent to science that wish to correct ingrained malpractices or counter the restrictions of a too far-reaching patent law. The following study addresses the changed legal conception of public access to information (II), legal guarantees of the freedom to publish (III), and the special protection afforded scientific research in patent law (IV).

II. The Changed Relationship between Secrecy and Public Access to Information

Modern societies are largely based on public access to information and the transparency of the individual spheres of society. We have recently seen trends toward greater freedom of information that would have been inconceivable just two decades ago. I would like to illustrate these trends first by discussing two representative fields.

1. Freedom of Information in Administrative Law

Public access to administrative records demonstrates the changes that have come about in a short period of time in efforts to provide the necessary degree of secrecy and the necessary degree of informational freedom. Data and official secrecy has long been a central feature of the German administrative tradition. It was the publication, in 1990, of the European Community Directive on the Freedom of Access to Information on the Environment (Directive 90/313/EEC of 7 June 1990, OJ 1990 L 158/56) that ushered in a period of profound change. The principle of public access to data (at least to the environmental data of the authorities) has been in effect since then, granting everyone a right to access administrative information. This transition has continued to the present day: a number of Germany's federal states have passed administrative information laws that confer upon citizens the right to access to all records kept by the authorities. These laws are modeled on those in other European countries, particularly in northern Europe.

We need not go into detail here. What is important is the changed conception of the law: official secrecy and the nondisclosure of data, once ironclad pillars of statehood, have now been overturned. This does not mean that state or corporate secrets, or the privacy needs of third parties, are neglected in all cases, yet they are no longer blanket concepts that can be used to create a sphere of secrecy. They must be elaborated and weighed against opposing public or private publication interests (*Scherzberg* 2000: 289). In this process, relevant legislation provides weighting regulations based on the principle of public access. This procedure corresponds to the provisions set out for the organs of the European Community in Art. 255 of the EC Treaty.

2. Access to Information in Corporate Law

Of greater interest to our study is the trend toward enhanced freedom of information and transparency in commercial enterprises, particularly in listed corporations. The information at issue here is often highly important for corporate policy and for the valuation of companies. In this way it is comparable to the data from cooperative industrial research projects. Despite its sensitivity, there has been a move toward substantially greater disclosure practices. Influenced by EC legislation and pressured by international financial markets, German law has also undergone significant change. What is notable is the great variety of information that must be reported: not only the distribution of company ownership and internal facts that affect stock price, but also personal information about company officers. Today even the board members' salaries are no longer taboo (*Lutter* 2003: 740).

In essence, the reporting requirements concern data that was once regarded as top secret. Today one hardly encounters arguments favoring secrecy in interpretations of the relevant regulations in literature on commercial law. In this area, too, a profound change in the conception of public access has been effected in an astonishingly short period of time. The laws relating to science and its institutions have a long way to go to catch up.

3. Comparable Arguments

The debates over access to files and access to scientific research have been conducted separately up to now. With good reason: they are not easy to conflate. Our primary purpose in addressing access to both administrative records and corporate information has been to demonstrate how quickly and effectively former bastions of secrecy can be conquered. However, it is also of interest to note that the functional arguments for public access to administrative information are similar to those found in the literature on the sociology of science (*Scherzberg* 2000: 202 ff.): freedom of information enables the public to monitor processes and formulate criticism. It has a two-sided, recursive effect. On the one hand, it improves the quality of public discourse, on the other, it intensifies the administration's perception of its environment. "By strengthening the communicative competency of society, the administration enhances its own informational resources, ensuring its acceptance by society and increasing its effectiveness." Further: "The administration's exposure to a critical evaluation by the public can be the source of insight into the necessity of early and exhaustive cost-benefit studies for individual social planning measures. It can also ensure access to both subjective states and confusions concerning the growing uncertainties inherent to the risk society. These subjective states can have an adverse effect on the populace's quality of life and opportunities for development even if individual risk prognoses are not confirmed and the self-referential rationality of the administrative system gives no grounds for intervention." The conclusion drawn here is that opening up public discourse leads to an administrative culture that is better tailored to citizen needs and that contributes to the social reintegration of the political-administrative system. One repeatedly encounters similar arguments, based on communication, transparency, trust and a willingness to learn. Companies and administrations have long since betaken themselves to the "Agora" (to use a phrase by *Nowotny et al.* 2001: 183 ff. and 201 ff.), and it is time for science to follow.

III. Publication of Research Findings: Freedoms and Obligations

In German law, legal problems relating to science are generally addressed at first as *questions of basic rights*. This is also true of publication issues: publication is itself perceived as an element of academic freedom, and commercial use of scientific results is considered an element of the property guarantees and of occupational freedom. Germany's Basic Law (*Grundgesetz* [GG]) deals with these issues in Art. 5, para. 3, as well as in Art. 14 and 12.

1. The Double Significance of the Constitutional Guarantee of Academic Freedom

Under Art. 5, para. 3 of the GG, art, scholarship, research and teaching are free. Freedom of teaching does not release any person from allegiance to the constitution. Otherwise, the text mentions no other restrictions; in particular, it does not contain the option appended to most of the other constitutional guarantees, that of restricting the protected sphere of such freedoms by Acts of Parliament or other sub-constitutional law. However, we must not take this to mean that the constitution guarantees unlimited academic freedom (the opposite will be shown).

Art. 5, para. 3 of the GG grants individual scientists the right to pursue their research and teaching with as few restrictions as possible. At the same time, it designates, as a constitutional value, a system of science that is structured as liberally as possible. Art. 5, para. 3 of the GG not only protects science from state intervention, but also calls for state measures to implement the idea of free research (Bundesverfassungsgericht [1973] E 35, 114). These may be necessary to resolve conflicts of interest between science and other sub-systems of society, or between researchers and research institutions within the scientific community. The constitutional guarantee of academic freedom is thus no guarantee of an absolute freedom, nor is it an unregulated freedom. It is a constituted freedom (*Kirchhof* 1986). We must differentiate between academic freedom as a political demand and academic freedom as a constitutional guarantee. Three regulatory perspectives must be distinguished for the *constituted freedom*:

- First, certain circumstances may permit, or even oblige, a state to intervene in academic freedom in order *to protect third parties*, or to safeguard key public interests. Scientific interests do not take precedence in every case. However, any concrete restriction of academic freedom can only be justified by reasons set forth in the constitution itself. These can take the form of the basic rights of third persons such as property or health protection, or state goals such as environmental protection (*Dickert* 1991: 411 ff.). General expectations, fears and desires, as well as political opportunism, are inadequate reasons for intervention. "To foster academic freedom, one must constantly bear in mind the concept that underlies it, namely, that the findings of a science that is liberated from the idea of social benefit or political expedience best serve the state and society." (Bundesverfassungsgericht [1978] E 47, 370).

- Second, state regulations are often necessary to make possible scientific communication and to organize scientific processes in the first place. Science is in many ways dependent on state institutionalization. Art. 5, para. 3 of the GG can thus be regarded as a fundamental organizational right that is subject to constitutional restrictions but also allows for the necessary institutionalization of science (see *Trute* 1994: 173 ff.; *Kleindiek* 1998: 239 ff.). In these cases, it calls for state regulations to create an appropriate balance between various interest groups. No single interest has a schematic priority here. Such a priority rule can be ruled out for the very reason that conflicts may arise between two legal entities that are equal in every respect. One need only think of two researchers on the same research team who are unable to agree as to whether their results are worthy of publishing! By the same token, there is no fixed ranking of the individuals or institutions enjoying academic freedom.
- Finally, state regulations may be necessary *to protect free research* from the actions of others. In this case, Art. 5 para. 3 of the GG functions as a regulatory mandate. For instance, one might ask whether and to what extent the state can, or even must, protect a scientist's intentions to publish from the commissioning party's right to secrecy. In this context, patent law is a good example. It is primarily the fundamental rights of third parties (including the intellectual property rights under Art. 12 and 14 of the GG) that place limits on state regulations designed to advance science.

Two things become apparent when we examine the interplay between constitutional guarantees on the one hand and the regulatory mandate for state legislation, for the judiciary and for the authorities on the other: the constitution recognizes science as a relatively autonomous area clearly demarcated from the other sub-systems of society. However, it also assumes that there are diverse interrelations between, and a mutual permeation of, these sub-systems. Totally isolated science no more adheres to a constitutional principle than does its total merger with politics, industry and the media. Science's behavioral patterns, standards, and norms are important for the guarantees contained in the constitution (*M. Blankenagel* 2001: 81 ff.); yet they are not identical with these guarantees. From the perspective of the constitution, the failures of science to attain to its ideal self-conception are not pathological, but they do warrant observation.

It is primarily the task of parliamentary legislation to balance out the different requirements of social sub-systems. This process takes place in the public eye, often supported by hearings with experts or by special parliamentary commissions of inquiry. The executive does not possess its own interventionary mandate; however, if it proceeds in a "science-friendly" fashion and employs specifically scientific procedures, it has more freedom in organizational design. The courts – particularly the constitutional court – exercise a control function over the legislative and the executive. As a result, academic freedom is not subject to despotic state intervention but safeguarded not only by the public nature of the parliamentary procedure, but also by administrative procedures amenable to science and by legally defined control standards. The constitutional guarantee of academic freedom forms a reflective framework in which the processes of political negotiation and judicial control can be observed. This encourages all those involved to make comprehensible assessment decisions and strive for science-friendly regulations.

2. Publication Obligations versus Publication Freedoms

The issue of publication also illustrates the double significance of Art. 5, para. 3 of the GG as both a subjective *Abwehrrecht* (defensive right) and the objective designation of free research as a constitutional value worthy of protection. Since science may be seen as a context of action and communication, its protection must not be confined to safeguarding individual volitional decisions (*Trute* 1994: 245 ff.). The protection of science as a system must also be guaranteed (*A. Blankenagel* 1980: 69).

a) Point of Departure: The "Positive" Freedom of Publication

In the prevailing view, Art. 5, para. 3 of the GG protects, from state bans, a researcher's freedom to publish: it makes publication restrictions impermissible as a matter of principle. This is the "positive" freedom of publication (*M. Blankenagel* 2001: 76 ff.). Restrictions are only justifiable if they are necessary to protect constitutional goods of equal value, including the right to privacy and the protection of state secrets. Data privacy laws specify a number of publication restrictions that can be seen as generally accepted. Property protection is also regarded as a reasonable justification for imposing limitations (short-term, at least) on the freedom to publish. However, general political considerations or diffuse attempts at political correctness do not entitle a state to suppress the results of scientific research.

b) Problem Area: The "Negative" Freedom of Publication

We will not further explore the conflicts linked to the positive freedom to publish. They do not go to the heart of the above-mentioned developments in publication practice that are critical for the scientific system. These problematic developments have not resulted from restrictions imposed by the state on the freedom to publish. Rather, they stem from researchers who refrain from publishing materials expected of them or who agree to publish only with the consent of those commissioning their work. In the traditional view, Art. 5, para. 3 of the GG also protects the right of scientists to keep their findings to themselves or to make them known to a limited circle of persons only. This is the "negative" freedom of publication (see *Burzik* 1996: 47 ff.). Against this backdrop we can see how the frontlines have shifted: the potential for conflict is not generated by state intervention, but by secrecy clauses in research contracts, license agreements and grant awards. The press repeatedly reports on such procedures, particularly in pharmaceuticals research. The issues raised here are different from those surrounding publication restrictions by the state, but they are nevertheless bound up with the guarantee of academic freedom and its double significance as an individual right and objective constitutional value:

- Should not Art. 5, para. 3 of the GG be interpreted as containing an *obligation* to publish?
- If this proves infeasible, should we not at least exclude the negative freedom of publication from the protective domain of academic freedom?
- What powers and obligations does Art. 5, para. 3 of the GG grant the legislature, the judiciary and the administration to counter privately agreed-upon informational restrictions that prove deleterious to science?

c) An Obligation to Publish as a Requirement Directly Derived from the Constitution?

It would be easy to answer these questions if it were possible to derive a publication obligation directly from the constitutional guarantee of academic freedom. However, such an assumption of a *fundamental obligation* is extremely difficult to substantiate. Art. 5, para. 3 – which appears under the heading *Grundrechte* (Basic Rights) in the constitution – makes no reference to obligation. The constitutions of other European states, including those of Spain and Portugal, are not as reticent. Here we find references to "basic rights and duties." The European Charter of Fundamental Rights has even incorporated the concept of obligation into its preamble: "Enjoyment of these rights entails responsibilities and duties with regard to other persons, to the human community and to future generations."

Yet even the *Grundgesetz* specifies a few individual duties in connection with basic rights. It expressly mentions the social obligations inherent to property: its use "shall serve the public good" (Art. 14, para. 2, GG). Likewise, Art. 6 of the GG states that care and upbringing of children are not only the parents' right, but a duty primarily incumbent on them. Even more interesting for our context is the compulsory schooling – as an obligation imposed by a culturally defined state *(Kulturstaat)* – that is said to be included in Art. 7, para. 1 and 5 of the GG. Ensuring public access to research results can also be counted among the vital interests of a *Kulturstaat*. Publication is the precondition for the validation of findings by other scientists; the permanent availability of published material is a fundamental requirement for a knowledge-based society. These two facts show that public access to information covers more ground than academic freedom seeks to protect through individual *Abwehrrechte*. So there are definitely ways of reconciling both the freedom and the obligation to publish on the basis of constitutional law.

However, the strongest arguments against the assumption of a direct, constitutionally anchored obligation to publish are drawn not from theoretical but from practical legal factors. Basic obligations are *leges imperfectae* that require concretization in law. Only the legislature can define these duties with the required precision to make them applicable to all designated cases. This concretization is a vital necessity for the issue of publication because different regulations must be drawn up for university, extra-university and industrial research, dealing with (among other things) the strength of the obligations and the manner in which they are to be fulfilled. Furthermore, legislators would have to ensure that research results considered inconclusive by scientists are not subject to these obligations.

However, in a few areas – particularly in fundamental research projects funded by the state – we are currently seeing the emergence of clearly defined obligations. When the courts interpret regulations and contractual clauses, they may avail themselves of these obligations without being hampered by the requirement for legal concretization.

d) Exclusion of the Negative Freedom of Publication

Others have proposed excluding the negative freedom of publication from the protected domain of academic freedom. This idea is not quite as far-reaching as the assumption of a direct constitutional obligation to publish. However, it plays a similar role in facilitating public access to science: research conducted by scientists who intend from the start to keep their results secret should not enjoy the protection of Art. 5, para. 3 of the GG. Those who refuse to provide details of their research findings are no more members of the 'scientific community' than those who refuse to take note of other researchers' germane findings (*Dickert* 1991: 277 ff.; *M. Blankenagel* 2001: 73 ff.). This is a persuasive point. The exclusion of the negative freedom of publication reduces the previous protective shield and makes it easier for the legislature to impose publication obligations for sound reasons of the public good, without being encumbered by the high standards incorporated into Art. 5, para. 3 of the GG. Researchers will also be subject to indirect pressure to publish their findings if the so-called experimental use exception (found in environmental, tax and other branches of sub-constitutional law) is deemed as being inapplicable to research work not intended for publication.

This proposal has the appeal of an options model that gives individual scientists the freedom to use a protective regime of their choosing for their research. Even so, objections remain: if academic freedom is meant to safeguard creativity and innovation, then it should not be permissible to exclude, with such ease, research not intended for publication from this protected sphere. It is definitely possible for this research to contribute partially or indirectly to the generation of knowledge – at least within the research institution that has conducted it.

3. Art. 5, Para. 3 of the GG as Creating an Objective Constitutional Value: The State Obligation to Protect Free Research

These findings suggest the feasibility of defining these problems as those of sub-constitutional law, primarily of statutory law. It is worth noting that Art. 5, para. 3 of the GG includes much more than an *Abwehrrecht* defending against state actions. Academic freedom also calls for active protection by the state itself. In contrast to most of the other basic freedoms, this feature was not appended later on but an integral component of the guarantee of free research from the start (*A. Blankenagel* 1980: 40). Consequently, it is not an addition that strengthens an individual *Abwehrrecht*, but a fundamental decision that seeks to accommodate the key function of free research in the overall development of society. According to the vivid formulation of the Federal Constitutional Court (*Bundesverfassungsgericht*), it encompasses " ... the responsibility assumed by any state that conceives itself as a *Kulturstaat* for the idea of free scholarship, as well as state participation in its implementation. It obliges the state to orient its actions to this goal in a positive sense, that is, by introducing protective and incentive measures that will prevent the erosion of the guarantee of freedom." (Bundesverfassungsgericht [1973] E 35, 114).

In terms of legal theory, the objective component is important in two ways: first, criteria can be derived from it to justify the imposition of restrictions on opposing legal positions, even those *within* Art. 5, para. 3 of the GG. This constitutional value has not only a legitimizing but a motivating function. It can become so urgent to protect the interests of science that a state is not only authorized, but obliged to act. Up to now, this semantic level has primarily played a role in the creation of adequate structures for research institutions. Two other problems are crucial for the present context:

- First, it may become necessary to restrict the (negative) publication freedom of individual researchers in order to ensure the public-access interests of free research.
- Second, it may become necessary to safeguard researchers' free access to sources of information by restricting intellectual property rights, etc., that are too broadly defined (see IV).

In both cases, the argument of an objective constitutional value is made to confine the scope of individual rights. Seen thus, academic freedom has much in common with the freedom of broadcasting guaranteed in Art. 5, para. 1 and para. 3 of the GG. These address not only the protection of a freedom but also the protection of an institution and its functions. It is of vital importance for *all* basic informational rights in Art. 5 of the GG that due consideration be given to the context of individual actions, without which their interpretation is impossible.

4. A Plea for "Positive Regulation" of the Publication System

The issue of research publication, in particular, shows that any legal examination that confines itself to the defensive character of basic rights is far too limited in scope. The statement issued by the Federal Constitutional Court on threats to free broadcasting is also relevant to the research system: the threat of misuse emanates not only from the state but from other forces in society. Therefore, it would be mistaken to perceive Art. 5 of the GG as merely containing an *Abwehrrecht* defending against state actions (Bundesverfassungsgericht [1994] E 90, 88). Academic freedom also requires a "positive regulation" by the state if its action principles and its objectives are not to be misused for commercial or political purposes. Here we also see the validity of the *democratic principle* (and, it must be added, the republican principle of a true "*res publica*"), which plays an important role for all types of freedom of information. To avoid any misunderstanding: science does not always act according to democratic rules. Questions of methodology are not settled by majority resolutions, nor are research results obtained on the basis of voting regulations. Yet academic freedom in a democratic society is dependent on transparent actions and public access to research results. The law can be expected to provide a legal framework for this. For instance, lawmakers might consider imposing an obligation to publish research results on all *publicly funded research* – regardless of whether the funding goes to institutions or to specific projects. This measure could be supported by a ban on awarding third parties a right to determine, through publication decisions, the way results are ultimately to be used.

In the long run, regulations of this nature would also change publication practices in *industrial research*, which is beyond the reach of state funding. It is important to recall that industrial research and industrially commissioned research work are no longer synonymous with secrecy. Examining the model contracts that public research institutions are endeavoring to use as a legal foundation for cooperative research ventures with industry, we can make out a trend toward taking publication issues much more seriously. It is rare to find provisions that grant the commissioning party an absolute veto right over publication of research results. Clauses are common that permit a commissioning party to review a patent application without being undermined by prepublication of results (which would make it impossible to satisfy innovation requirements). For this purpose, periods of sixty or ninety days are not uncommon. But otherwise these contracts are founded on the principle of publication.

Of course, what is decisive are not the model contracts, but the contracts that find acceptance in industry. The universities are the primary source of concern here. Many have insufficient experience and are swayed by arguments that the funds will be channeled to other universities in the event that requests are not met. The problem is particularly acute in dealings with large corporations. The more comprehensive regulation of the publication system, as proposed in this paper, seeks to strengthen the position of research institutions in contractual negotiations by showing them the limits of what can be negotiated.

In view of the material content of a publication obligation that is derived directly from the constitution, it is above all the *courts* that should devote greater attention to the idea of public access to information in their interpretation of research contracts. Contractual clauses that generally prohibit the publication of research results or that link publication to an absolute veto right should only be accepted if there are concrete reasons for protection of state secrets or personal privacy. Military and corporate secrets can be accommodated quite effectively in this fashion too. As a whole, however, laws relating to science and scientific organizations should define standards that are in no way inferior to those currently recognized by commercial and corporate law.

However, in those cases where secrecy and partial publication of findings purposely create or maintain false impressions, publication restrictions should generally be rejected. Courts should not hesitate to treat transgressive contractual clauses as null and void under § 138 of the German Civil Code (*Bürgerliches Gesetzbuch*). Here, a more stringent interpretation of the law would primarily serve to strengthen scientific publication interests in contractual negotiations, among other fields.

5. The Complementary Significance of Scientific Self-Regulation

This leads us to another important point. Regulation by the state can only partially resolve the problems that science has "with itself" since it must confine itself to fundamental principles. This means that the "positive regulation" implied by Art. 5, para. 3 of the GG is largely dependent on scientific *self-regulation.*

Scientific codes of conduct (guidelines, recommendations) as well as other rules of good scientific practice are not legal regulations per se. Some deal with more technical issues of the research process. Some contain statements that are vital for the self-conception of science and reflect its normativity and ethos. These codes may play an extraordinarily important role in the practical work of individual researchers. Examples include guidelines for good laboratory practice and publication rules detailed by the editorial boards of professional journals. Germany was late in discovering the importance of fully elaborated scientific standards. The "Proposals for Safeguarding Good Scientific Practice," passed by the German Research Foundation in 1998 to deter research malpractice, represent a milestone (*Schmidt-Aßmann* 1998). Today, all major scientific institutions have similar guidelines. One finds, in the USA, codes of conduct touching upon an even larger number of scientific self-regulation issues. These cannot be enforced by legal means but are sanctioned by social and, above all, internal scientific mechanisms, including the loss of the scientist's reputation within the scientific community.

These guidelines only acquire legal relevance when they are referred to in laws and incorporated into relevant legal obligations. "Adoption clauses" of this kind can be found in employment contracts or the staff regulations of research institutions. They are also alluded to in some statutory regulations – for instance, when approval of new pharmaceutical products hinges upon observance of the rules of "good laboratory practice." It is only the framing of laws based on such guidelines that will determine the extent to which the behavioral rationality of science is reflected in the legal order of the state. Conversely, scientific standards liberate public law from its required regulatory tasks. These standards are more closely oriented to scientific practice and are more effective as experimental regulations. It is possible to develop a functionally differentiated, complementary relationship between public law and scientific standards.

Science must also develop its own code of conduct for publication issues, particularly for the problem of publication bias. This applies primarily to professional associations and the editors of scientific journals, but also to research-funding agencies as well as the research institutions themselves. The following rules would appear necessary (among others):

- Rules that require disclosure of a commercial relationship between the author and third parties who have a financial interest in the research object.
- Rules that require researchers to mention the incompleteness of their findings, including the use of unverified research data supplied by commissioning parties.

It should be possible to develop a typological system of rules for such conflicts of interest that is in no way inferior to the regulations designed to eliminate bias in trial and administrative procedural law.

IV. Access to Science and the Significance of Patent Law

The public character of science is also shown by the permanent accessibility of scientific knowledge. Like the obligation to publish, the perception of acquired knowledge as a public good is an integral component of the idea of free science. However, there are also counter-arguments – some legitimate and readily comprehensible, others weak and unacceptable. Here, the law also has the responsibility to reconcile conflicting interests while giving due consideration to the behavioral rationality specific to science.

We will examine these issues using the example of patent law, which, by awarding specific proprietary rights, creates a regulatory framework for access to information that is designed to provide constant incentive to generate new knowledge. Instead of specific details, we will focus on structural issues: the constitutional basis of patent law (1), the central meaning of patentability (2), and the role played by the research clauses used by patent law to promote research progress (3).

1. The Constitutional Basis of Intellectual Property Law

In the prevailing view, copyrights and patent rights, as assets, enjoy the protection of the constitution's property guarantee (*Fechner* 1999: 152 ff. and 186 ff.).

a) Approach Based on Individual Rights

As early as 1971, the Federal Constitutional Court emphasized the relevance of this view to copyright law. Shortly afterward, it also confirmed its application to patent law: by awarding and safeguarding rights of ownership, use and disposition, Article 14 is assigned the function, within the overall framework of the constitution, of granting a proprietary freedom and enabling individuals to develop and shape their lives in autonomous fashion. According to the court, this protective and defensive function of the property guarantee requires us to view the author's powers over the assets created by his work as "property" in the sense of Art. 14 of the GG (Bundesverfassungsgericht [1971] E 31, 239). The Federal Constitutional Court has also addressed the question of general rights for inventors, stating that such rights have long been regarded as a subject of the law and, as technical copyrights, must also be categorized as property (Art. 14 of the GG). The European Charter of Fundamental Rights is even clearer: intellectual property is protected under Art. 17, para. 2.

Thus, at the outset, we encounter a position clearly favoring individual rights. The Constitutional Court does not regard copyrights and patents as a public good but interprets the constitution as providing fundamental legal protection to originators. There is no way of getting around this fact on the basis of constitutional law. "In the legal order, the right to commercially exploit a new idea that advances technology and science is granted to the person who has produced this idea." (Bundesverfassungsgericht [1974] E 36, 290). As a matter of principle, the legislature is required not only to grant originators the assets resulting from their creative acts, but also to ensure their freedom in using these assets as they deem fit.

b) Drafting Legislation with a Private and Public Benefit

Despite their categorization as property under the constitution, copyrights and patent rights are not predefined to the same degree as laws relating to real estate. The Federal Constitutional Court correctly points out the necessity of drawing up more detailed regulations that make copyright and patent law practicable in the first place (Bundesverfassungsgericht [1988] E 79, 40). This is a task of the legislature, which must base its work on the private utilization of such rights on the one hand and the public good on the other (Art. 14, para. 2 of the GG). The originator's individual interests cannot claim absolute priority. In intellectual property law, in particular, we are witnessing a growing interest in allowing comprehensive public access to intellectual creations since they form the basis of scientific and cultural production. Taking the principle of proportionality and the necessity of equal treatment into account, lawmakers must bring these contradictory interests into a balanced relationship while also considering publication and access-to-information goals.

2. The Limits of Patentability

These goals are supported by the definition in § 1, para. 2 of the Patentgesetz (Patent Law [PatG]), which states that discoveries, scientific theories, mathematical methods and data-processing programs are non-patentable. Indirectly related to this are regulations forbidding the issuance of patents for discoveries whose use or publication threatens the public order or violates morality. It is here that the public order meets the public domain (*Barton* 2004).

It is not surprising to find major disputes raging in this area between groups interested in preserving public access to information and others intent on expanding private disposition rights. The issue of patentability determines the extent to which the diagnosed "propertization of scientific data" can progress. From a legal viewpoint, such problems involve interpreting and, if necessary, altering relevant national, European and international legislation. Posing the question in this way illustrates not only who the agents are (courts, parliaments, international conferences), but also the arenas in which the various interests are negotiated. A knowledge of these circumstances makes it possible to discuss what steps need to be taken and what forces activated to produce specific effects favoring public access. Here the law performs a structuring task: it pinpoints the places where political deliberations must begin.

Jurisprudence faces two important decisions here – first, in the field of biopatents and second, in the question of the patentability of computer-implemented inventions. Conflicts in both areas are reflected in the directives drawn up by the European Union to "harmonize" the national patent laws of its member states. Two positions have been staked out: 1.) the patent-friendly position adopted by the EC Commission and the governments cooperating in the Council to increase competitive capacity in international markets; and 2.) the restrictive position adopted by the European Parliament and the parliaments of many member states. In both cases, the most important conflicts in research policy can be traced back to concrete formulations in the directives:

- The Community Directive on the Legal Protection of Biotechnological Inventions (Directive 98/44/EC of 6 July 1998, OJ 1998 L 213/13) deals with the question of whether there should be absolute legal protection for chemical substances in gene sequences. The answer to this question determines the extent to which the holder of an original patent can control the subsequent development of the patent's diverse functions. The directive favors providing absolute legal protection for chemical compounds, even though the special features of biological substances do not make this choice an obvious one. Nonetheless, the directive does stipulate that the commercial application of a gene sequence must be described concretely in the patent application. Critics of the directive have argued that the individual regulations are not precise enough and that they generally reveal a wavering attitude (*Kraßer* 2004: 225 ff.; *Keukenschrijver* in: *Busse* 2003: § 1a, note 12). In such cases, the bodies applying the law – that is, patent offices and patent courts – have the last word. Normally, the judicial development of a law is considered advantageous because it can proceed experimentally and flexibly. However, fundamental patent questions such as these show that the framework of the individual patent procedure is not broad enough to accommodate the various interests before a decision is taken. It is the European or national lawmakers who must make fundamental decisions in this field.

– The release of the Directive on the Patentability of Computer-Implemented Inventions (Proposal for a Directive COM (2002) 92 final of 20 February 2002) has caused similar structural difficulties. This proposal by the European Commission requires member states to view computer-implemented inventions as "belonging to a technological field." This would lead to a sweeping, positive resolution of the dispute over the patentability of software. In contrast, the European Parliament has issued a statement insisting "... that data-processing is not regarded as a technological field in the sense of patent law, and that data-processing innovations are not viewed as inventions in the sense of patent law either." In its "Common Position," the Council, as the third legislative body of the EC, opted not to include a corresponding rule in the directive. But the Council's explanatory statement concerning the proposal reflects the Commission's campaign for a broader interpretation of patentability. The issuance procedure for the directive has not yet been concluded. If the Council and the Parliament are unable to reach an agreement, the proposed directive will not be issued, in accordance with Art. 251 of the EC Treaty. There is reason to believe that the Conciliation Committee will negotiate further formulations that will do more to obscure than to resolve the fundamental conflict. This is another instance where the power to decide fundamental questions will be shifted (back) to the patent offices and patent courts.

The scientific community need not accept, without complaint, the developments in patent law currently underway or foreseeable for the future. However, if undesirable developments are to be avoided or rolled back, the community will have to mobilize political forces.

3. Research Privileges in Patent Law

An additional structural problem – in this case, the international interconnectedness of patent law – can be studied on the basis of the so-called experimental use exception, a legal construct specially designed to reconcile the originator's constitutionally recognized rights of use on the one hand and constitutionally protected scientific and cultural interests on the other. Corresponding regulations can be found in national and international patent law. Under § 11, no. 2 of the PatG, patent protection shall not extend to "acts done for experimental purposes relating to the subject matter of the patented invention." A clause with nearly identical wording can be found in the European Patent Conventions of 1975 and 1989 (Art. 27 b). This clause contains the specific mandate to protect free research that is incumbent upon any state that defines itself "as a *Kulturstaat*" (Bundesverfassungsgericht, [1973] E 35, 114). The experimental use exception is closely linked to the central patent law goal of providing incentive to produce new knowledge through restricted access to information (*Kraßer* 2004: 813; *Osterrieth* 2004: note 255). In this field, legal developments in Europe and the USA are proceeding in entirely different directions: whereas Germany, in particular, has seen liberal rulings, American courts are increasingly reining in the experimental use exception.

a) Recent Developments in German Law

Through a broader interpretation of § 11, no. 2 of the PatG, the courts are rightly seeking to accommodate the increased importance of research clauses in the scientific community. Older interpretations only permitted experiments to be conducted that served scientific purposes and that were designed to test the operability and functionality of a patented invention. However, the Federal Court of Justice (*Bundesgerichtshof*) set an entirely new course with its ruling of 11 July 1995 ("Clinical Trials I"; Bundesgerichtshof [1995] BGHZ 130, 259). The court broadly defined the concept of experimentation to include any methodical procedure for the acquisition of knowledge, independent of the use to which the acquired knowledge was ultimately put. According to the court, it was of no significance whether the experiments were conducted to verify patent specifications or to obtain more advanced results. Nor was the possible pursuance of other goals decisive, even commercial ones. The Federal Court of Justice drew this conclusion from the purposes of patent protection. On the one hand, patent law endeavors to recognize special achievements by awarding exclusive rights; on the other, it intends to spur on additional developments. According to the court, the foremost duty of patent law is "to encourage technological progress and to profitably promote the spirit of invention for industry." The court made express reference to the freedom of research in

Art. 5, para. 3 of the GG, which, it argued, must be given due consideration when the patentee's interest in comprehensive legal protection is weighed against the general public's interest in ongoing technological development. This ruling places the general public's interests on nearly an equal footing with the patentee's in terms of legal protection. In a 1997 ruling, the Federal Court of Justice not only continued this course but defined it more clearly.

Although the new interpretation of § 11, no. 2 of the PatG has placed a substantial burden on the patentee, we can make no objections to it on the basis of constitutional law. The Federal Constitutional Court refused to consider a complaint of unconstitutionality on this issue, explaining that research clauses are permissible legislative instruments for promoting the research and the development of science and technology (Federal Constitutional Court, 1st Chamber of the 1st Senate, GRUR 2001, 43). The Constitutional Court also stated that the Federal Court of Justice's application of § 11, no. 2 of the PatG is compatible with Art. 14 of the GG, because – against the backdrop of freedom of research and the social obligations inherent to property – it was comprehensibly shown that the unrestricted protection of a patent is not justified if it hinders further development of a technology. With this ruling, the developmental policy objectives of recent adjudication may be considered accepted. Nor did the Federal Constitutional Court recognize, as an unacceptable burden on the holder of the product patent, the danger that the new research will rapidly lead to the issuance of use patents. The holder of the product patent ultimately profits from the commercial value of the use patent. This once again underscores the significance of the incentive function of patent law.

b) Comparative Legal Observations

Comparative law plays a special role in the field of commercial proprietary rights. In its recent rulings, the Federal Court of Justice explicitly refers to the European Patent Convention of 1975/1989 and draws comparisons to rulings in other contracting states. An analysis of British and Dutch adjudication also reveals a tendency to no longer restrict the research clauses of their respective laws to experiments that serve solely scientific purposes (*Cornish* 1998: 737 ff.). Legal scholars point out that these rulings accommodate new research conditions in academic-industrial cooperative ventures, in which it is impossible to isolate purely scientific research goals. The more recent, broader interpretation of the experimental use exception can thus be seen as a convincing functional answer to changed research conditions. A new path has been taken that does not call into question the fundamental value of patent law and which integrates industrial patentees into the developmental mandate. Insofar as their position is restricted in favor of the continuance of progress, they too are forced to make their contribution.

The developments in Europe stand in marked contrast to those in the USA. In its ruling of 3 October 2002 in "Madey v. Duke University," the United Court of Appeals for the Federal Circuit states that the experimental use defense must be very narrowly defined in patent violation procedures. It must be confined to actions "solely for amusement, to satisfy idle curiosity, or for strictly philosophical inquiry." Actions are impermissible that serve clearly recognizable or primarily commercial purposes, or that have even slight commercial repercussions. These standards are meant to apply to universities as well. Their legal form as non-profit educational institutions does not automatically mean they qualify for the experimental use exception. Many scholars have asked whether "Madey v. Duke University" is not an attempt to respond to the universities' own, increasingly expansive patenting practices. Criticism of the court's narrow interpretation will probably predominate as a matter of fact. However, there are differing opinions as to how great the obstacles will be to new research as a result of the ruling (see *Guttag* 2003 and *Pate* 2002: 254 ff.). All things considered, the discussion shows that research clauses are integrated into a network of instruments and procedures in patent law and patenting policy. If, for example, it is easier to get compulsory licenses, there is less reason for a broader interpretation of the experimental use exception.

4. Outlook

We have sought to illustrate the crucial role that patent law plays for the topic of this paper by examining two problems linked to its *legal structure*. A broader treatment of this subject would have to examine a variety of related questions: to what extent should publicly financed research exploit the potential of patent law to protect findings and to make specific use of licenses for research policy? What repercussions will an orientation to the economic utilization of research results have for both the research process itself and the self-conception of science? These are questions that not only require practical and political reflection, but must also consider (or call for) a legal framework. All told, law plays a substantial role in the problem of intellectual property rights since these rights are themselves legal constructs. Even slight changes in the law can have far-reaching consequences for the open access to information in the future. As a 2003 publication of the Royal Society aptly puts it: "Laws that are drafted thoughtfully and applied wisely can encourage innovation, reward creators and entrepreneurs, and promote economic and social gain without leading to unacceptable monopolies or unduly restricting freedoms." (The Royal Society 2003: 29).

Bibliography

Barton T (2004) Der "Ordre public" als Grenze der Biopatentierung. Erich Schmidt, Berlin

Blankenagel A (1980) Wissenschaftsfreiheit aus der Sicht der Wissenschaftssoziologie. Archiv des öffentlichen Rechts (AöR) 105: 35–78

Blankenagel M (2001) Wissenschaft zwischen Information und Geheimhaltung. Pro Universitate, Baden-Baden

Burzik M (1996) Legal Aspects of Communication in Scientific Research in the FRG. In: Andrea Orsi Battaglini (ed) Freedom of Information and Confidentiality in Scientific Communication. Nomos, Baden-Baden, pp 11–142

Busse R (2003) Patentgesetz, 6th ed. De Gruyter, Berlin

Cornish WR (1998) Experimental Use of Patented Inventions in European Community States. International Review of Industrial Property and Copyright Law (IIC): 735–753

Dickert T (1991) Naturwissenschaften und Forschungsfreiheit. Duncker & Humblot, Berlin

Fechner F (1999) Geistiges Eigentum und Verfassung. Mohr Siebeck, Tübingen

Groß T (1992) Die Autonomie der Wissenschaft im europäischen Rechtsvergleich. Nomos, Baden-Baden

Guttag EW (2003) Immunizing University Research from Patent Infringement: The Implications of Madey v. Duke. Journal of the Association of University Technology Managers (AUTM) vol XV: 1 ff.

Kirchhof P (1986) Wissenschaft in verfaßter Freiheit. Decker & Müller, Heidelberg

Kleindiek (1998) Wissenschaft und Freiheit in der Risikogesellschaft. Duncker & Humblot, Berlin

Kraßer R (2004) Patentrecht, 5th ed. C. H. Beck, München

Lutter M (2003) Corporate Governance und ihre aktuellen Probleme, vor allem: Vorstandsvergütung und ihre Schranken. Zeitschrift für Wirtschaftsrecht und Insolvenzpraxis (ZIP)24 I: 737-743

Nowotny H, Scott P, Gibbons M (2001) Re-Thinking science: knowledge and the public in an age of uncertainty. Polity Press, Oxford

Osterrieth C (2004) Patentrecht, 2nd ed. Beck, München

Pate GN (2002) Analysis of the Experimental Use Exception. North Carolina Journal of Law & Technology (N.C.J.L. & Tech.) 3, Issue 2: 254 ff.

The Royal Society (2003) Keeping science open: the effects of intellectual property policy on the conduct of science (available under http://www.royalsoc.ac.uk/files/statfi)

Scherzberg A (2000) Die Öffentlichkeit der Verwaltung. Nomos, Baden-Baden

Schmidt-Aßmann E (1998) Fehlverhalten in der Forschung – Reaktionen des Rechts. Neue Zeitschrift für Verwaltungsrecht (NVwZ)17: 1225–1234

Trute HH (1994) Die Forschung zwischen grundrechtlicher Freiheit und staatlicher Institutionalisierung. Mohr, Tübingen

Weingart P (2001) Die Stunde der Wahrheit? Velbrück Wissenschaft, Weilerswist

The Response of the Legal Order: Final Commentary from a Social Studies of Science Perspective

Helga Nowotny

To speak about science and the law often conjures up images of science in court – the role of scientists as expert witnesses or images of murder trials and the tricky, albeit serious questions of evidence that need to be weighted and assessed. A whole new array of forensic methods and techniques, like DNA analysis, that utilize the latest state-of-the-art scientific knowledge and know-how for the purpose of rendering justice, continue to fascinate the media, while underlining the dependence of law upon scientific expertise. Yet, this picture is at best partial.

At another level, however, a different form of interchange and interaction marks the mutual interdependencies of science and the law and their respective relationship to the State. At the origin of modern politics these were the two institutions that had assumed the task of curbing the arbitrariness of political power and the arbitrary will of political agents. Science did so by gradually replacing religion in certifying the reality of the natural world. The laws that governed it were beyond the reach of political power. Therefore, science could position itself at the service of explaining and manipulating Nature and, by doing so, was able to assume the stance of an apolitical authority (*Ezrahi* 2003). The law had its own struggles with the political authorities of the time and continued to do so over time. It too had to strive towards a position of apolitical authority if it was to set limits to the political powers in a secular state.

Yaron Ezrahi argues that the contemporary shift from modernity to post-modernity in the sphere of politics today has involved a significant erosion of the ability of scientific knowledge and of scientific institutions to continue to constitute such a source of apolitical authority. The role of science and technology as apolitical authority rested upon a cluster of normative perspectives with science often functioning as a surrogate extension of first-order political authority. This authority constitutes and upholds the overall normative and synoptic perspective which in turn sets the boundaries of second-order politics. This is the recognized sphere of the political, in which the contest over material and symbolic resources takes place. According to Ezrahi the comprehensiveness and stability of such an overall normative perspective, regardless of whether it is hierarchical or egalitarian, is more important for the apolitical status of science within the political order than its specific content (*Ezrahi* 2003: 68). Now we are witnessing the partial disintegration of a hegemonic normative system. The systematic, comprehensive and relatively coherent normative frame which was centered around the state perspective on collective goals and public policies, is giving way. To the extent that contemporary political configurations no longer rest as firmly as they did on the idea of the polity as a relatively coherent whole or the individual as a stable discernible unit, the expectation that science would provide an incontestably neutral perspective on issues of public policy can no longer be sustained and science looses much of its former power as a political resource in the justification of state policies. It becomes depoliticized in the sense of loosing its basis in the first-order political authority. Science must now continually negotiate its claims to represent neutral and objective knowledge by means of a variety of institutional, intellectual and rhetorical strategies.

This is where, potentially and actually, the normative force of the law comes into play. If the diagnosis is correct – and there is broad agreement on it reached through different approaches and perspectives – the law is now called upon to compensate for the loss of the former authority of science. It is important to realize that this authority, while apolitical in its outward appearance, was political because it was derived from and sustaining the political order of the State. If the public nature of science is under assault today, the assumption is that the capability of the law to protect the public interest is still intact and that, due to its normative force, the law continues to possess the necessary capability and flexibility to regulate and shape the conditions under which scientific knowledge production occurs. The law is now called upon to protect the political authority of science as being apolitical, capable of producing objective knowledge and to an extent that remains to be regulated, above private and conflicting interests.

This is where the constitution comes in as the highest legal authority. This is also where the insistence of the law enters to keep the sub-systems of society differentiated and hence separate from each other. This is why the law must insist that it is – legally – possible to distinguish political issues and questions from economic and scientific ones. This is why the normative force of the law must be counter-factual and disregard empirical evidence to the contrary. And yet, at the end of the day the law must prove itself to be sufficiently flexible and capable of the necessary adaptions. More concretely, the law is called upon to meet the challenges that arise out of a dual transformation that the science system undergoes at present. While the overall transformation of the political order, which severely impact the political authority of science, is largely out of the control of the science system, since it is a derived authority, the science system can respond and shape its own response to the dual transformation in which it is so deeply involved. It is at this level where science and the law meet in different ways and in different areas. At the level of the constitution, science is safeguarded and protected by the highest authority of the State. At the level of day-to-day encounters and conflicts, the law provides one of several resources that science can use to its advantage or that will be used to its perceived disadvantage. At this level, science has to meet new demands and expectations and negotiate its claims.

The challenge consists in how to meet a dual transformation. One transformation makes itself felt primarily within the science system itself, although its consequences are far-reaching and will in the long run determine the extent to which science can maintain its nature as a public good. Due to its undeniable success and achievements the science system continues to expand, necessitating a much stronger financial investment which can no longer be met by public funding alone. To the extent that science is perceived as the motor of economic growth, strong economic pressures are being exerted upon the production of scientific knowledge. The expectation is that the production of scientific knowledge will bring about economic benefits in the short- and medium-term and that it will enhance economic competitiveness. One strategy towards achieving this goal lies in fostering closer ties between academic research and industry. Among other, this has led to a convergence of what previously were two distinct cultures. By reconfiguring the previous ethos of an economically disinterested science with the ethos of industry and the business world, scientists are urged to become research-entrepreneurs. In the eyes of the public, this renders the alleged neutrality of science further questionable. Indeed, the boundaries that before have been vigorously defended by science in its rhetoric, become blurred. Science and the law are called upon to respond to this situation. What is public and what is private need to be re-defined together with the complex issues of Intellectual Property Rights. It is obvious that the outcome will weigh heavily on how society will fare in guaranteeing the indispensable, although always relative, autonomy that science will need also in the future.

The second transformation is also brought about through the overall success of science and technology and the enormous impact on society and the lives of citizens exerted thereby. Although the recent scientific and technological advances have brought about overwhelmingly beneficial results, they have also been accompanied by newly emerging risks. By offering many more choices with partly new underlying dilemmas the range of uncertainties and of perceived risks has also been extended. The relationship between science and society in the past decades has therefore been marked by previously unknown tensions and mutual distrust. Science has experienced not only a decline of its former cognitive and social authority, but often an outright refusal from a part of society of exactly those products and 'deliverables' in which it takes great pride of achievement. On the side of society the feeling of being by-passed and ignored in its most articulated disquietudes and concerns about risk is pervasive. Frontier research meets increasingly with suspicion and troubled concerns where it touches the sense of self and what it means to be human.

Public controversies about real or perceived risks have led to demands for greater participation in the political and scientific decision-making processes pertaining to complex scientific-technical issues. In the context of these debates, the concept of "scientific citizenship" has gained ascendancy. If it is to evolve from being mere rhetoric or symbolic politics, its content and form will need to be worked out more concretely at the interface of science and the law. Based on the assumption that rights and duties exist not only on the side of citizens, but equally on the side of public institutions, the political-administrative institutions as well as the scientific institutions are both challenged to step out from the shadow of the State and to become more actively and openly engaged in their interaction with citizens. The demands for greater accountability, for open access to information and for 'democratizing (scientific) expertise' have been raised and are unlikely to disappear. They challenge the law and legal scholars to come up with innovative solutions that can provide guidance and regulate the conditions under which these demands can be met and, perhaps, future conflicts be mediated in the framework of democratic politics. The contributions in this volume offer promising avenues of how to move forward. They also contain clear warning signs for further trouble ahead. Schulze-Fielitz, for instance, in his typology of conflicts, focuses on the thorny issue of how contemporary democracies will deal with the request of minorities who, on faith-based or other, value-related grounds, refuse any compromise (*Schulze-Fielitz*, this volume).

The three contributions to this volume coming from legal scholars in the fields of constitutional and administrative law offer fascinating insights into the responses of the legal order to the challenges raised by the dual transformation. They cover a wide range of the arenas in which science and the law meet. Some are marked by public controversies about risks, while others take place with scientists in their research labs or institutions as the main protagonists. Time and place differ – and both matter. These encounters allow the reader to catch a glimpse of the processes that show 'law in the making'. Law is never a-historic, even if the normative framework to which it is bound and which it seeks to impose is of a long-term and only slowly changing nature. Principles that constitute the guiding overall normative framework are therefore laid down in the constitution. It is one of the primary goals of the law to guarantee that expectations addressed to the future are secure, meaning that they will be met in specific ways that are foreseeable already at present. Nevertheless, the law must be capable of evolving in response to overall changing societal conditions and, more concretely, in response to newly arising demands for regulation. The legal order may do so sometimes with delay. At other times the law, in a progressive mode, may even anticipate societal developments. The legal contributions in this volume provide a sense for the caution with which the law proceeds, but they equally make clear that the caution is guided by principles and guidelines that are stipulated in the constitution. Thus, the authority of the first-order politics becomes linked to the second-order where arrangements need to be worked out in concrete settings and be implemented accordingly.

Place also matters. Overwhelmingly, it is still the nation-state which provides the frame, structure and procedures in which the legal order manifests and expresses itself (in this case, German law). However, the legal framework of the European Union makes itself felt increasingly. Sometimes, this translates into a pioneering role for EU legislation. Unencumbered by national tradition and precedent, EU legislation can take a more advanced position on some issues, evidenced, for instance, in environmental law. In other areas, the law of the European Union is under obligation to harmonize legal provisions that exist in member states and in some instances, like in the efforts to 'democratize expertise' the alleged gap in the democratic legitimacy of the European Union puts it under additional pressure to demonstrate democratic openness.

So, what can be observed on the various construction sites where the responses of the legal order translate into law in the making? Which are the salient issues that come to the fore and how do legal scholars go about in meeting them? They are, to put it succinctly, well informed about ongoing developments within the science system and in the relationship between 'science' and 'society'. This includes familiarity with the literature coming from the field of STS, science and technology studies. But the observation, description and analysis offered by these approaches are taken up in a highly selective way. If interdisciplinarity occurs through familiarizing oneself with the "ways of knowing" practiced in other disciplines, the interdisciplinarity of the law is constrained. Legal scholars have no problem in their understanding of constructs and the processes through which they come into existence, but they are adamant in according priority to legal constructs. While the observations and analysis provided by scholars of history or social studies of science may provide useful background information that finds its way into legal reasoning, the use that legal scholars make of it is strongly conditional and constrained by the legal world view. There seems to be not only no alternative to the legal point of view – the temptation to see the world differently does not even arise.

Legal scholars are aware, to take one example, that rights and duties are themselves legal constructs (*Schmidt-Assmann*, this volume). They know that however slight the changes in the law may appear to the outside, the consequences may be far-reaching and of enormous practical efficacy. They are aware that procedures matter and need careful design. In other words, there is a powerful and coherent system of concepts, procedures, institutions, and tools at their disposition through which they encounter the world of science and research, of modern technologies and their impact, and the conflictual or preventive interaction between the State, Science, Markets and Politics that is in need of regulation. If the intersections between law and other sub-systems of society can be approached at all through some form of interdisciplinarity – they can and they should – one must nevertheless be aware that the interdisciplinarity involved is of a special kind. It resembles the role that mathematics plays for the natural sciences. It connects different domains of knowledge and enables practical action, yet the concepts and tools through which this occurs retain a remarkable autonomy.

There is a clear grasp of deeper issues and how they connect. When addressing the problem of 'democratizing (scientific) expertise' for instance, this does not mean that everybody is an expert now. But efforts to translate such a political demand into legal procedure begin with the acknowledgement that scientific expertise has become a matter of public concern and that new demands for accountability have arisen that do not exempt science as an institution or grant immunity to scientific expertise. The selection of experts therefore becomes crucial as does the identification of stakeholders in the actual process of decision-making. The range of possible 'solutions' is seen to reside with the selection procedures of experts. Procedural rules can be extended further to the debates of experts, their ways of reaching agreements or how dissenting views should be incorporated. Here, the law and the legal scholars that make it, can be observed in constructing a public space – a contemporary agora – with the purpose of providing a framework for accountable political decision-making in matters pertaining to complex scientific-technological issues.

In approaching the contemporary transformations, in particular the increasing pressure towards commodification of scientific knowledge and the propertization of science, the tendency of the political receding and becoming replaced by 'governance', as Dominique Pestre maintains, cannot be overlooked. Technoscience, to use his terminology, is in grave danger to be at the mercy of the new power of a financial markets-driven regime. What can the legal order do to counteract it? Hans-Heinrich Trute's commentary highlights in a sharp and precise way the difference in approaches. While the analysis coming from social studies and history of science strongly underline the tendencies of the sub-systems to dedifferentiate and towards becoming increasingly blurred, the law, by contrast, maintains a strong preference for clear demarcations. These are clearly stated and endorsed at the level of the constitution. They can also be found at lower levels and the recurrent insistence that it is not only possible, but necessary, to distinguish and separate issues and questions pertaining to the legal, the political or the scientific realm. Only such demarcations and clear separations of sub-systems, the legal argument goes, provide sufficient safeguards against the intrusion of 'external' influences. Merging or overlapping interests that do not respect existing boundaries may provoke counter-action which in turn may lead to the further erosion of public trust or to other, negative results.

Such a position should first of all be taken at what it is: a clear endorsement of the public character of science endorsed and protected through the (German) constitution. It provides the anchoring point through which the State opts for a well-functioning science system, respects its public nature and guarantees its autonomy. The law provides, as Hans-Heinrich Trute convincly argues, sufficient flexibility for adaptations, whether they assume the form of opening up free access to information that previously was regarded as the State's prerogative for secrecy or making room for the plurality of scientific expertise. The important point is that the law (and the State) do not waver in their normative aims and objectives.

But – and this is the question that arises from a social studies of science perspective – can the legal order formulate and implement normative criteria that will be sufficiently robust to withstand the assault on the public nature of science that the contributions in this volume analyze? As already stated, legal thought in the making is well informed about actual developments, the main tendencies are seen and taken into account. But the normative stance adopted by the legal order is itself the result of a long historical process that has evolved at the origin of modern politics as a way of curbing the arbitrariness of the political power, but soon entered into a strong alliance with the State. Will the normative ideal of an autonomous science prove itself to be sufficiently resilient also at the present historical junction in curbing the voracity of markets for appropriating science as a public good for private, largely economic, gains? And if, as Hans-Heinrich Trute also suggests, many of the protests that now seem to be directed at the science system, are merely triggered by it and need to be re-directed to the political system which is the real target, what can the law do in order to encourage politics to be more responsive and inclusive?

The fact that empirical studies suggest a blurring of boundaries between State and the Market, between science and markets, between markets and politics, in no ways invalidates a normative stand which, indeed, is counter-factual. Nor are empirical and historical studies of the science system devoid of taking a normative stand themselves, which is strongly underlined by Dominique Pestre's contribution to this volume. The remaining question is simply whether the normative ideal enshrined in the law is sufficiently robust for a future that will differ enormously from the past when the ideal separation – as well as the intense cooperation – between State, Science and Industry was the norm.

A new divide between the public nature of science and its private, or privately owned, side in the production of scientific knowledge, has opened up. Likewise, the fact has to be acknowledged that public participation in decision-making in matters of science and technology, even if it may seem naïve as a panacea to the public's worries, will not disappear and has to be reckoned with. When science and the law, at the origin of modern politics, each succeeded after long struggles to constitute themselves as apolitical and neutral authorities, standing above particular interests and serving the public good, their (relative) autonomy was autonomy from political interference and their authority could be used to curb the arbitrariness of political power.

However, they could so only because the objectivity of science was also serving the State, and because, as Ezrahi has shown, the authority of science was closely linked to the first order authority of the State. The problem for science lies in solving a paradox: can science be part of the political process and yet separate?

One answer that has been given after the Second World War was the model of science conceptualized in Vannevar Bush's highly influential *Science: The Endless Frontier*. The model he proposed rested on the distinction between basic and applied research (itself, a distinction that goes back to the second half of the 19th century). Basic research was characterized simply as the unrestrained quest for knowledge, driven by curiosity and hence 'pure'. But Bush cleverly turned basic research into something that was also fundamental to societal progress and hence became part of the political process. Science was both separate from, and yet a part of, politics and decision-making. This conceptual model worked remarkably well and, despite many signs to the contrary, still has its adherents today (*Pielke Jr.* 2005). However, the problem with it is that the term basic research has become meaningless for many of the most advanced research fields in the life sciences and the environmental sciences have become deeply embroiled in their struggle to be part of the policy process and yet separate from it.

Today, science as a public good is being redefined. Political power, at least as power of the State, has partly given way to the pervasive influence of markets. State boundaries, while still existing, are overlaid by transnational agreements and trade. Globalization continues at a relentless pace. Science can no longer base its authority on the claim of its special relationship to Truth or of speaking in the name of Nature. Both these claims have receded in significance and have been replaced by instrumental values and the efficacy of science and technology to contribute to wealth creation and economic competitiveness. What matters today above all are the ties to industry and to markets, to the production of technical artefacts and tangible benefits to society. Arguably, this can be achieved in a public and private mode, but the consequences will strongly differ. In the end, we are back to an elementary normative question: what kind of society do we want to live in? Whatever the answer, science and the law will be crucial in determining the outcome. They ought to continue to learn from each other by speaking to each other.

References

Bush V (1945) Science, the Endless Frontier. A Report to the President. U.S. Government Printing Office

Ezrahi Y (2003) Science and The Postmodern Shift in Contemporary Democracies. In: Joerges B, Nowotny H (eds) Social Studies of Science and Technology: Looking Back Ahead. Kluwer Academic Publishers, Doordrecht, pp 63–75

Pestre D (2005) The Technosciences between Markets, Social Worries and the Political: How to Imagine a Better Future? In this volume: 29

Pielke Jr. R (2005) review of "Nature's Experts: Science, Politics, and the Environment". Nature 434: 130–140.

Schmidt-Aßmann E (2005) Free Access to Research Findings and Its Limitations. In this volume: 109

Schulze-Fielitz H (2005) Responses of the Legal Order to the Loss of Trust in Science. In this volume: 63

Trute H (2005) Comment from a Legal Perspective. In this volume: 53

Index

About the Authors

Helga Nowotny is currently Chair of EURAB, the European Research Advisory Board, advising the European Commission on research policy. Until recently, she has been Professor of Philosophy and Social Studies of Science at ETH Zurich, Director of Collegium Helveticum and Founding Director of Society in Science: The Branco Weiss Fellowship based at ETH Zurich. She has published widely in the field of social studies of science and technology, while maintaining an active interest in science policy. Formerly she was also Professor at the University of Vienna and Permanent Fellow of Collegium Budapest/Institute of Advanced Study. She was a Fellow at the Wissenschaftskolleg in its first year, 1981/82. – Address: Wissenschaftszentrum Wien, Strozzigasse 10/16, A-1080 Vienna, Austria. email: helga.nowotny@wzw.at

Dominique Pestre studied physics and history. He is currently Professor of History of Science at Ecole des Hautes Etudes en Sciences Sociales and Director of the Centre Alexandre Koyré in Paris. He has worked extensively on the history of 20th Century physics. Recently, he co-authored, with Amy Dahan, *Les Sciences pour la guerre, 1940-1960.* He also published on the historiography of science and on philosophical questions. One of his recent books in this field is *Heinrich Herz, l'administration de la preuve*, 2002. His current field of interest is the changes which affected research practices over the last decades. See his book, *Science, argent et politique, un essai d'interprétation*, 2003. – Address: Centre Alexandre Koyré, Pavillon Chevreul, Museum National d'Histoire Naturelle, 57 rue Cuvier, 75005 Paris, France. email: pestre@wanadoo.fr

Eberhard Schmidt-Aßmann is Professor of Public Law, Director of the Institute for German and European Administrative Law and Member of the Berlin-Brandenburgische Akademie der Wissenschaften. His general research interest is the theory of administrative law and especially the guidance and control of societal processes by state and administrative law. His concept of administrative law is enfolded in *Das allgemeine Verwaltungsrecht als Ordnungsidee* (2. Aufl. 2004) [General Administrative Law as an Ordering Idea). The law of science is one of his fields of reference and he published on the interplay between state law and scientific standards and self regulation. He was for several years referee of the German Research Association (DFG) and Scientific Member of the German Research Council (1993-1999) – Address: Juristische Fakultät, Friedrich-Ebert-Anlage 6-10, 69117 Heidelberg. email: e.schmidt-assmann@urz.uni-heidelberg.de

Helmuth Schulze-Fielitz studied law and social sciences. Currently he is Professor of Public Law, Environmental Law and Public Administration Sciences at the Bayerische Julius-Maximilians Universität Würzburg. He works on the theory and practise of constitutional and administrative law and public administration sciences. Among others he published *Der informale Verfassungsstaat,* 1984 (The informal constitutional state), *Theorie und Praxis parlamentarischer Gesetzgebung*, 1988 (Theory and Practise of Parliamentary Law Making), *Was macht die Qualität öffentlich-rechtlicher Forschung aus?* in: Jahrbuch des Öffentlichen Rechts der Gegenwart 50 (2002) (What makes the Quality of Research in Public Law). – Address: Juristische Fakultät, Julius-Maximilians-Universität Würzburg, Domerschulstraße 16, 97070 Würzburg. email: schulze-fielitz@t-online.de

Hans-Heinrich Trute studied law and social sciences. He is currently Professor of Public Law, Media and Telecommunication Law at the University of Hamburg. Since 2001 he is Director of the Centre "Law and Innovation" at the Law Faculty of the University of Hamburg and Member of the Research Institute for Public Administration at the University of Speyer. From 1999 until 2005 he was Ombudsman for scientific misconduct of the German Research Association (DFG). His main focus is the theory and development of administrative law and public administration sciences. He also published widely on the interface between law and science. The concept is laid down in his book *Die Forschung zwischen grundrechtlicher Freiheit und staatlicher Institutionalisierung* (Research between Freedom and Institutionalisation by the State), 1994. – Address: Universität Hamburg, Fakultät für Rechtswissenschaft, Edmund-Siemers-Allee 1, 20146 Hamburg. email: hans-heinrich.trute@jura.uni-hamburg.de

Printed in the United States
91329LV00003B/187-198/A